D0948467

SPATIAL COMPONENTS OF MANUFACTURING CHANGE, 1950–1960

by

Yehoshua S. Cohen
The Hebrew University, Jerusalem

and

Brian J. L. Berry
The University of Chicago

THE UNIVERSITY OF CHICAGO
DEPARTMENT OF GEOGRAPHY
RESEARCH PAPER NO. 172

1975

Published 1975 by The Department of Geography
The University of Chicago, Chicago, Illinois

Library of Congress Cataloging in Publication Data

Cohen, Yehoshua S 1937–

Spatial Components of Manufacturing Change, 1950–1960.
(Research papers—University of Chicago, Department of Geography; 172)
Bibliography: p. 259

1. Industries, Location of—United States—History. 2. Space in economics. I. Berry, Brian Joe Lobley, 1934– joint author. II. Title. III. Series: Chicago. University. Dept. of Geography. Research papers; 172.

H31.G514 no. 172 [HC110.D5] 338'.0973
75-33654
ISBN 0-89065-079-9

Research Papers are available from:

The University of Chicago
Department of Geography
5828 S. University Avenue
Chicago, Illinois 60637
Price: $6.00 list; $5.00 series subscription

PREFACE

As the results of the latest Census of Manufacturers have been published in the past year, the findings have been greeted with both amazement and alarm. The old cities of the Nation's traditional manufacturing belt heartland now have rapidly declining industrial bases as industry has decentralized within their urban fields, diffused to smaller urban centers and dispersed into what were formerly hinterland regions. The Mid-South has emerged as a new manufacturing belt with industries located in small towns and rural areas, rather than centralizing in industrial metropoli. But there should have been neither amazement nor alarm, for those trends were already in evidence to the discerning eye during the 1950's, as rapid changes in accessibility transformed the traditional calculus of economic advantage. This is a study of those changes during the 1950's. It was begun several years ago--long before the latest census data were collected--and completion was delayed somewhat by sickness. The delay may have been opportune, however, because the findings serve as a counter to hasty generalization based upon rapid inspection of the more recent statistics.

The support of the Economic Development Adminstration during the early stages of the study is acknowledged, as is subsequent assistance from the Center for Urban Studies of the University of Chicago and the University of Illinois at Chicago Circle.

Chicago August, 1975

TABLE OF CONTENTS

CHAPTER 1

INTRODUCTION

The long-term trend for manufacturing industries to _disperse_ from the old manufacturing belt on a national scale, to _diffuse_ from larger to smaller urban centers within the nation's urban system, and to _decentralize_ within urban regions, has been observed at least for the last half century (Fuchs, 1962; Creamer, 1969). Yet despite growing concern for the consequences of an apparent acceleration of these trends in the past two decades, they have yet to be subjected to a rigorous spatial analysis on a consistent nationwide basis. Such is the purpose of this study.

We begin by reviewing the more important background literature. This is followed by a discussion of the spatial concepts used to structure the analysis and their operational measurement. Next, the particular procedures of multivariate analysis of variance used to apply these concepts to measures of manufacturing are exemplified. Then follows the body of analysis. Finally, the detailed nuances of the decentralization and diffusion dynamics are codified, and a general overview of the changing spatial organization of American industry is provided.

BACKGROUND LITERATURE

Perhaps the most comprehensive recent attempt to describe trends in manufacturing location is the series of studies undertaken by The National Industrial Conference Board (NICB) that culminated in a 1969 report by Creamer. Creamer asked whether locational trends observed in the immediate post-war period, 1947-1958, had persisted into the mid-sixties, and he showed that there was both continuity and subtle change.

Consistent with earlier NICB investigations, he used seven locational types to structure his inquiry:

A. The principal city of an industrial area.

B. A satellite city with 100,000 or more population in an industrial area.

C. Remainder of the industrial area.

D. A city of 100,000 or more population outside an industrial area.

E. Remainder of the county in which a D city is located.

F. Important industrial county outside industrial areas, having 10,000 manufacturing employees but no city as large as 100,000 population.

G. Remainder of the United States.

Creamer argued that these categories represent different degrees of locational concentration of manufacturing employment, with A representing the highest concentration and G the lowest. Shifts from categories A and B to C were termed "primary diffusion" and correspond approximately to our term <u>decentralization</u>. Shifts to D, E and F were termed by Creamer "secondary diffusion" and are roughly similar to the concept of <u>diffusion</u> to be used later in our analysis. Finally, shifts to G were termed "dispersion;" however, such shifts are but a component of <u>dispersion</u> as we shall use that term. As will be noted later, Creamer observed total changes taking place in a classification of areas, whereas we view decentralization, diffusion and dispersion as processes that affect all areas simultaneously.

The principal finding of the Conference Board's 1947-1958 studies had been that, nationwide, a ten percentage point decline had taken place in the share of manufacturing located in the principal cities (the loss was 364,000 jobs in only four years, 1954-1958), accompanied by an offsetting rise in the manufacturing share of localities suburban to these cities through primary diffusion (decentralization), while only minor changes took place in those years in the shares located in the areas of secondary diffusion and in the vast hinterland of industrial dispersion.

This national pattern was made up in large measure of shifts taking place <u>within</u> the old manufacturing belt. Declining central city shares, growth in suburbia, and a standstill beyond were the essential characteristics of manufacturing change within the industrial heartland. Some small regional shift from the old industrial belt to other parts of the nation could be discerned, while outside the industrial heartland a somewhat different pattern was found. In the nation's hinterland regions (again, consistent with prewar changes), there were rising manufacturing shares in the principal cities as well as in their suburbs, no change in the areas of secondary diffusion, but a decline in the counties of dispersion.

These trends were most apparent in factory employment, and were partially offset in the principal industrial cities by the high concentration and rapid growth of central administrative offices and auxiliaries, enhancing the continuing role of the major cities as leaders in research and development and innovators of new activities.

Creamer's analysis was of what happened during the 1958-1963 recessionary phase of the national economy, and in the expansionary period that followed, 1963-1966. For the 1958-1963 period, Creamer reported that some manufacturing shifts were consistent with the earlier trends, but that some important differences had appeared as well. The share of manufacturing employment in the old manufacturing belt continued to drop, from just over to just less than two-thirds of the national total. Principal cities in the industrial areas continued to lose jobs - 338,000 in all. Growing shares were found in the suburbs - 433,000 jobs - as well as in large cities outside industrial counties (F), and also in the countryside, (G), beyond. The earlier regional difference in trends between the nation's manufacturing-belt heartland and outlying hinterlands was eliminated as hinterland changes came increasingly into conformity with heartland shifts, producing a consistent nationwide pattern. Central administrative offices and auxiliaries, while still experiencing absolute growth in the principal cities, underwent a sharp decline in share (from 61% in 1954 to 52.1% in 1958 to 45% in 1963) in favor of suburban locations. On the other hand, the principal cities (A), satellites (B) and suburbs (C) of the industrial areas still retained 80% of central administrative employment and 77% of research employment in 1963, with a reverse orientation of the two: two-thirds of the central administrative offices were located in the central cities and two-thirds of the research and development laboratories in the suburbs. Thus, established industrial regions _in total_ retained their dominance of innovative capabilities.

Exactly the same trends were discovered by Creamer between 1963 and 1966, during the period of accelerated economic expansion. No longer were marked regional differences evident, according to Creamer, and redistribution on a broad regional basis from the old heartland to the hinterland was minimal. The principal losses were again to be found in the major industrial cities, while expansion took place throughout the zones of primary and secondary diffusion, and beyond, in the areas of dispersion. Additional new data on creation of new establishments, liquidation or migration of existing ones, and growth in the existing cohort of establishments, permitted Creamer to make additional generalizations for this period, too: The principal cities of industrial areas (A) remained as important incubators of new establishments, consistent with the dominance of existing industrial regions in the central office, auxiliary, and research and development functions, although the gains in employment from this source were over-shadowed by high liquidation rates of both new and old production activities, and migration of many production industries to other areas, while the continuing cohort of industries grew at an average pace. Migration from the principal cities was the principal source of growth in the largely suburban areas of primary diffusion (C), along with very rapid growth of the continuing cohort. Finally, growth in zones of secondary diffusion (D, E and F) was largely due to existing cohort expansions.

THE NATION'S ECONOMIC SPACE: STRUCTURING DIMENSIONS AND LOCATION PROCESSES

There is a strong suggestion in the foregoing that a nationwide *inter*regional equilibrium in relative manufacturing growth was achieved during the 1960's, while in Creamer's terms, continuing primary and secondary diffusion proceeded *intra*regionally as smaller urban centers gained manufacturing employment relative to larger centers and the cores of larger urban regions lost relative to their suburbs. But Creamer's analysis presents several problems. Creamer began with a typology of areas, and he examined relative manufacturing growth rates within the framework of an *a priori* classification of types of change. For example, if growth rates in his type C areas exceeded those in types A and B this was, for him, sufficient to argue that "primary diffusion" was taking place. This kind of prior classificatory constraint is clearly inadequate to elicit in full the richness of the changes that were unfolding. Decentralization, diffusion and dispersion can and do work simultaneously. The suburbs of a middle-sized southern metropolitan area may, for example, be gaining industrial employment as a consequence of all three processes while the suburbs of a similar-sized metropolitan area in New England may be gaining due to decentralization, but losing both because industry is diffusing to smaller urban centers and because industry is moving from the older industrial heartland to newer industrial growth areas in the nation's hinterlands.

There is a need, therefore, to analyze industrial change as it unfolds simultaneously along each of the dimensions of a three-component space:

1. Relative accessibility within the heartland-hinterland structure of the national economy.
2. Relative status within the hierarchical structure of the nation's urban system.
3. Relative location within the city-suburb community structure of metropolitan regions.

To each corresponds a particular change process--dispersion, diffusion and decentralization -- and the total change in any area is some mix of the three. Such an analysis is what is attempted here.

Market Potentials and the Heartland-Hinterland Structure of the National Economy

The starting point in any such analysis of the nation's economic space must be Harris's (1954) suggestion that maximization of market potentials is a location factor of growing significance to American industry, and can be equated with profit maximization in many cases. In this view, if profits are maximized by maximizing market potentials and if market potentials change, so will market opportunities, and dispersion of industry to take advantage of the new opportunities will follow.

Market potential P at some point i, P_i, is, of course, normally computed as $\sum_j \frac{m_j}{f(d_{ij})}$, where m_j is the market "mass" at j and $f(d_{ij})$ measures the intervening distance frictions. In Harris's view, entrepreneurs seek out points of maximum accessibility within the most concentrated market area, thereby ensuring the best protection against future entry by competitors and achieving all-around cost reductions based on the economies of scale, service and advertising present in those markets which are the easiest to serve.

Until recently, no proof had been provided that Harris's speculations, in which he equated potential maximization with profit maximization, were indeed valid, although Harris's cartographic evidence suggested that his postulates were probably true. However, such a formulation was developed by Tideman in 1968. Tideman assumed that at any point o, the total cost function for production of a good is linear:

$$C_o = a_o + b_o Q_o. \tag{1}$$

Here C_o is total cost a point o, a_o and b_o are fixed cost and marginal cost respectively at point o, and Q_o is the quantity of output to be produced if o is the location of production. The marginal cost of output in the _ith_ market, b_{oi}, is given by

$$b_{oi} = b_o + r_{oi}, \tag{2}$$

where r_{oi} is the unit transportation cost from the production site o to the _ith_ market, which is assumed to be independent of the quantity shipped (i.e., marginal and average transport cost are equal). The addition to profits resulting from operations in the _ith_ market, π_{oi}, is

$$\pi_{oi} = Q_{oi}\left(P_{oi} - b_{oi}\right), \tag{3}$$

where Q_{oi} and P_{oi} are respectively quantity sold and the price in the _ith_ market when o is the location. Tideman also assumed that the demand schedule in each market has a constant price elasticity, η_i:

$$Q_{oi} = M_i P_{oi}^{\eta_i} \tag{4}$$

where M_i is whatever constant fits the equation.

Price theory says that profits in each market are maximized when marginal revenue in each market equals marginal cost in that market, that is, when

$$b_{oi} = P_{oi}\left[1+\frac{1}{\eta_i}\right] \qquad (5)$$

or

$$P_{oi} = \frac{b_{oi}\eta_i}{\eta_i + 1} \qquad (6)$$

Now (4) and (6) can be substituted back into (3) to obtain

$$\pi_{oi} = M_i\left[\frac{b_{oi}\eta_i}{\eta_i+1}\right]^{\eta_i}\left[\frac{b_{oi}\eta_i}{\eta_i+1} - b_{oi}\right], \qquad (7)$$

or

$$\pi_{oi} = \frac{M_i}{-\eta_i}\left[\frac{b_{oi}\eta_i}{\eta_i+1}\right]^{\eta_i+1} \qquad (8)$$

Defining

$$k_i = \frac{-\eta_i^{\;\eta_i}}{\left[\eta_i+1\right]^{\eta_i+1}} \qquad (9)$$

(8) becomes

$$\pi_{oi} = k_i M_i b_{oi}^{\;\eta_i+1} \qquad (10)$$

The significance of the parameter k is brought out by the following table. The numbers in parentheses indicate limiting cases.

$-\eta$	(1)	1.10	1.25	1.50	2.00	3.00	5.00	8.00	(00)
k	(1)	.714	.532	.382	.250	.148	.082	.049	(0)

Summing (10) over all markets and adding fixed cost back in yields the total profit associated with location at o:

$$\pi_o = -a_o + \sum_{i=1}^{N} k_i M_i b_{oi}^{\;\eta_i+1} \qquad (11)$$

If the assumptions of constant average transport cost, linear production cost, and constant elasticity demand curves are justified, and if a producer can set delivered price in each market separately, then (11) is a function of location whose maximization one might reasonably expect to explain location. Its inputs are the fixed and marginal costs at all possible producing points, a matrix of transport costs between the possible producing points and the N markets, and an elasticity of demand and a multiplicative constant to describe each market. As long as average transport costs are constant, first order approximations of the true production cost and demand functions (which exist in the assumed forms as long as the true functions have continuous derivatives) could be used to check whether a particular point was a local maximum of profits. If average transport costs are declining, there will be a "fixed cost" component to the transport cost of serving any market, and the markets served may vary with location. Price will be determined by the marginal transport cost, but the *ith* market will be served from location o only if π_{oi} is positive after deducting that part of transport cost not covered by marginal cost charges.

Some simplifying assumptions make (11) recognizable. In particular, assume that production cost and the elasticity of demand are the same everywhere. Then (11) becomes

$$\pi_o = -a + k \sum_{i=1}^{N} M_i b_{oi}^{\eta+1} \tag{12}$$

This is a (linear) monotonically increasing function of the summation, and therefore it is maximized when the summation is maximized. Recalling (2), the summation, denoted by T_o, may be written

$$T_o = \sum_{i=1}^{N} M_i \left(b+r_{oi}\right)^{\eta+1} \tag{13}$$

If M_i is identified as the relative size of the *ith* market (a reasonable interpretation of the multiplicative constants in a set of demand equations when they have the same elasticity), then the function T_o is like potential, except that a constant is added to transport cost, and the exponent to which this sum is raised is not necessarily -1. However, if marginal production cost is zero and $\eta = 2$, then (13) becomes

$$T_o = \sum_{i=1}^{N} M_i r_{oi}^{-1}, \tag{14}$$

which is exactly Harris's traditional expression for potential. It has not been shown that if changes in potential are directly proportional to changes in profits, then, necessarily, b=0 and η=-2, but it seems that any other result would require a highly combination of circumstances since (13) always varies in proportion to profits, within its assumptions.

If transport cost is zero, (13) has a value that is independent of the location of production. However, it is possible to derive a locationally dependent expression in the limit as transport cost goes to zero. Thus (13) can be rewritten

$$T_o = b^{\eta+1} \sum_{i=1}^{N} M_i \left[1 + \frac{r_{oi}}{b}\right]^{\eta+1} \tag{15}$$

which has a series expansion

$$T_o = b^{\eta+1} \sum_{i=1}^{N} M_i \left[1 + \left(\eta+1\right)\frac{r_{oi}}{b} + \left(\frac{\eta+1}{2!}\right)\frac{r_{oi}}{b}^2 + \ldots\right]. \tag{16}$$

In the limit as transport cost goes to zero, with a rate structure that is constant, the second and higher order terms become insignificant, and the remainder of the expression is

$$T_o = b^{\eta+1}\left[\sum_{i=1}^{N} M_i + \frac{\eta+1}{b}\sum_{i=1}^{N} M_i r_{oi}\right]. \tag{17}$$

The final summation is the only term that depends on o, and since $\eta+1$ is negative, T_o is maximized in this case when the final term, transport cost, is minimized.

Filtering Processes and the Hierarchical Structure of the Urban System

Other bodies of literature come to the fore when questions of manufacturing shifts within the urban system are raised, for example central-place theory and growth center concepts. But central-place theory, the better developed of the two, is static, and growth center concepts have yet to be linked satisfactorily to any well-developed theoretical argument.

One side of the growth center argument is that to optimize the impact of public investment on GNP, it is necessary to locate the investment in such a manner that the increment to GNP will be maximized. The argument continues that there is pursuasive evidence for strong effects of city size on productivity, of productivity on wages and incomes, and of incomes on the returns to migration. Thus, this argument concludes that a filtering strategy directed at small towns is likely to be counter-productive from the viewpoint of national welfare. To cite some of the evidence, William Alonso shows strong correlations of value added and city size in the U.S. (Table 1.1). One effect is seen differences by city size (Table 1.2), leading to income differences that persist even after deflated by regional cost of living differences (Figure 1.1). In consequence, there are strong positive inducements to migrate to larger cities (Table 1.3), a fact seen clearly enough in differential rates of population increase and migration by city-size class.

A countervailing feeling that these demonstrable advantages of large cities are coming unglued is expressed by such economic developers as Victor Roterus (1970):

> Over the period from 1958 to 1963 the lopsided concentration of industry in central locations and regions was being reversed. Whereas the Far West, as everyone knows, has been growing rapidly for a long time, the following three major regions of the country, notably laggard in the past, exceeded the national average of

TABLE 1.1

Value Added Related to City Size in 1963

Population	Payroll per Employee Index (1)	Value Added per Employee Index 2)	Value Added minus Payroll per Employee Index (3)	New Capital per Employee Index (4)
250,000 - 500,000	0.994	0.970	0.943	0.967
500,000 - 1,000,000	1.029	1.047	1.099	1.028
1,000,000 - 5,000,000	1.061	1.046	1.029	0.925
5,000,000 - and over	1.058	1.053	1.129	0.784

Source: Based on data in R. Douglas, "Selected Indices of Industrial Characteristics for U.S. Standard Metropolitan Statistical Areas, 1963," Discussion Paper No. 20, Regional Science Research Institute, Philadelphia, 1967.

TABLE 1.2

Ratio of Actual To "Expected" Hourly Earnings, by City Size, United States, 1959

	White Males	White Females	Non-white Males	Non-white Females
Rural	0.83	0.83	0.78	0.76
Urban Places				
Under 10,000	0.84	0.84	0.75	0.63
10,000 - 99,999	0.91	0.88	0.76	0.78
Standard Metropolitan Statistical Areas				
Under 250,000	0.97	0.94	0.84	0.76
250,000 - 499,999	0.97	0.96	1.04	0.90
500,000 - 999,999	1.02	1.03	1.07	1.00
1,000,000 - and over	1.12	1.13	1.10	1.19

Note: "Expected" hourly earnings were calculated for each sex and color by multiplying the national average hourly earnings of each color and sex by age and education by the annual hours worked by members of that cell in the region, summary in each case across all such cells in the region and dividing by the total man-hours of the region.

Source: Adapted from V. Fuchs, Differentials in Hourly Earnings by Region and City Size, National Bureau of Economic Research Occasional Paper No. 101, New York, 1967.

FIGURE 1.1

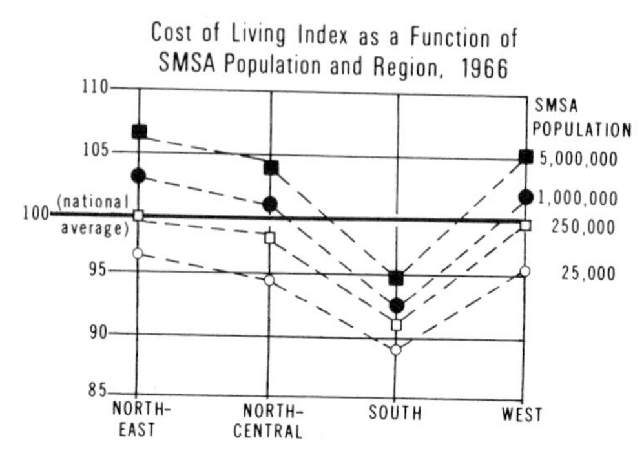

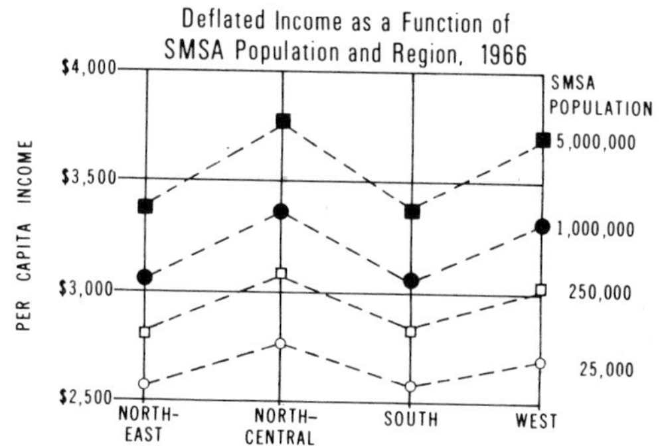

The upper chart compares the cost of a standard market basket betwee[n] areas and size of place, relative to the national average for the United States. Thus a person living in a metropolitan area of 5,000,000 in the Northeast would pay $1.06, while a person in an are[a] of 25,000 in the South would pay 88¢ of $1.00 worth of goods priced at the national average. The lower chart compares average incomes between areas and size of place after adjusting for cost of living differences given in the top chart. It can be seen that deflated incomes increase with size. It may well be that this represents additional compensation necessary to attract people to larger areas, though differences in average skill levels may also be involved. Other evidence available indicates the first explanation noted is of considerable importance. The data are obtained from regression equa- tions based on observations on forty-two areas derived from statisti[cs] compiled by the Bureau of Labor Statistics (U.S. Department of Labor and Office of Business Economics (U.S. Department of Commerce).

SOURCE: From 1970 Annual Report, Resources for the Future, Inc.

TABLE 1.3

Annual Earnings Gains of Migrants From the Rural South to the Urban North, Compared with Rural South Nonmigrants of Same Education, Race and Sex

	Years After Moving	TO SMALL NORTHERN CITIES[a] Elementary Education	College Education	TO VERY LARGE NORTHERN CITIES[b] Elementary Education	College Education
White Males	0-5	---	$3,075	$ 600	$3,075
	6-35	$1,550	2,175	2,700	3,700
Black Males	0-5	800	3,875	1,400	3,875
	6-35	1,550	2,175	2,000	3,000

a Under 50,000 population

b Over 750,000 population.

Source: From R. Wertheimer, The Monetary Rewards to Migration Within the U.S. Washington, D.C.: The Urban Institute, 1970.

manufacturing growth: Rocky Mountain, Southwest and Plains. The latter region, for example, exceeded the national growth rate for the first time, and by 29 percent.

More importantly, . . . not only have some of the lagging regions begun to exhibit definite growth tendencies, but the old trend of more and more concentration in metropolitan areas was definitely abating. Manufacturing employment grew at a rate of only 3.4 per cent in the metropolitan areas of the country between 1958 and 1963. On the other hand, the growth rate in the non-metropolitan areas of the country was four times greater than that -- 13.6 per cent compared to 3.4 per cent.

. . .

Why do industries go to small towns? The results of spot surveys can be summarized. Industries that have selected small town locations have given the following reasons for their decision: the desire to avoid congestion, opportunity to take advantage of untapped labor, attractive wage rates, availability of workers who can be easily trained, high productivity of labor, desire to avoid large city competition for labor, and availability of large plant sites at reasonable costs. It is noteworthy that five of these seven factors deal with different aspects of the labor reserve.

Other small town advantages frequently cited are more leisurely living, greater friendliness of small town people, existence of development groups ready and willing to do what they can to help an industry. Some of the companies reported that they wanted to become "part of a community" and this was considered easier to attain in a small town than in a large one. Some company management personnel felt that life was too short to spend so much time commuting to and from work.

. . .

Perhaps the most salient fact behind the apparent shift in industrial location is the improvement in the location capabilities of the smaller urban areas and corresponding deterioration in the merits of the big metropolitan area location. On the one hand, there has been a general improvement in most urban places in communications, transportation, supporting services and public facilities. On the other hand, the highly urban areas which in their hey-day offered the economies of agglomeration in the superior availability of services, markets, and skills now are becoming painfully aware of the diseconomies of congestion and high costs. More recently, there has been added what may become the most important factor of all -- what might be called the "turmoil factor" in our larger city areas.

. . .

Small towns can be roughly classified by their industrial development possibilities. Gustav Larson once categorized them as follows:

1. Small towns in the pale of large population centers. These towns will grow whether they want to or not. They will be favored not only by new industry from outside the region but by industry leaving the central city.

2. Administrative centers, university, and tourist towns should do well in the future because government, education, and tourism are growth industries. These towns are also generally pleasant towns to live in.

3. The last class, small towns in strictly rural areas, faces a most difficult job of area development.

Hence the roadblocks to growth include isolation from larger population centers, nonavailability of skills, and lack of services and facilities which larger industries need. While occasional large plants will locate in or near these rural-type small communities, the real hope for the future for these communities lies in the direction (1) of small industries based on local resources and markets, (2) of industries tapping local labor supply, and (3) of providing recreation, repair, and other services for nearby centers.

As if to extend some of Roterus' conclusions, the Appalachian Regional Commission has reported:

1. A selected set of Appalachian communities, ranging in size up to 28,000 population, and newly provided with highway access by either the Interstate Highway System or the Appalachian Development Highways had significantly accelerated manufacturing growth rates.

2. These communities also shared certain other common characteristics:

a) They had created industrial development agencies with authority to acquire large tracts of land and were developing them as industrial parks;

b) They had vocational education schools at the high school or post high-school level;

c) They were keeping ahead of their public service needs.

3. Otherwise similar communities equally provided with new access, but lacking a local "pool of leadership" did not develop.

If the trends described by Roterus are true, it becomes critical that they be explored in depth because research continues to point out that "changes in the level of manufacturing, particularly those associated with the development of new lead manufacturing sectors, has had the greatest single impact on the differential rates of growth of U.S. cities . . . the development of the manufacturing sector has provided the major thrust to metropolitan growth and the lack of development of the manufacturing sector has been the major hindrance to this growth" (Madden, 1970).

Recent contributions by Muth (1968) and Thompson (1968) become relevant at this point, marking a significant watershed in our knowledge of the economics of urban growth and the links between urban growth and regional development. These authors make a clear distinction between the processes affecting employment and incomes in those large cities (exceeding, say, a quarter of a million population) that have passed some threshold for self-sustaining growth, and those processes working in smaller places in which traditional export-base hypotheses might still obtain. Thompson, in addition, links the two in a "filtering down" theory of industrial location centering on diffusion processes working from level to level of the urban hierarchy.

Muth's elaborate analysis of the differential growth of large U.S. cities 1950-60 produced findings essentially supporting the Borts-Stein (1964) hypotheses concerning urban and regional development. Principally, his statistical investigations indicated that money wage rates are exogenously determined by firms selling their outputs and buying non-labor inputs in national markets. The fact that money wage changes were not affected by changes in employment, coupled with a low elasticity of the labor supply, suggested that effects of changes in the demand for a city's exportable sector output were quite different than those implied by the export-base hypothesis. The major effect of increases in exportable demand is to raise money wages, and such increases have relatively little effect on total employment, but merely change its sectoral composition, thus making it hard to sustain the hypothesis that, in large cities, employment change is exogenously determined. It is this change in sectoral composition in large cities that lies at the center of a "filtering" concept of industrial diffusion, however.

Continuing, Muth found that immigration of population into a large metropolitan area leads to as great a proportional increase in employment as it does in the civilian labor force, and that increases in employment in excess of the labor force growth that natural increase in population brings about are important stimuli to immigration. Therefore, a quantitatively important multiplier mechanism affects city growth, the value of which depends primarily upon the elasticity of migration with respect to employment. As a consequence, the base-service ratio has little or no effect in influencing differential total employment changes.

On the other hand, and quite surprisingly considering the abundance of evidence to the contrary, the hypothesis that differences in migration rates are a positive function of wage differences among the metropolitan areas was not borne out in his model equations. Therefore, he said, while it is fairly clear that there are advantages in migrating to the larger metropolitan areas, no differential economic advantage of migrating to any one in particular can be discerned. In consequence, as money wages rise, the tendency to increase aggregate money income in the city outweighs the substitution in production and consumption associated with the rise in money wages with the result that the proportion of the city's labor force employe in the domestic sector rises.

Thompson built on the notion, demonstrated by Muth, that no differential economic advantages accrue to large metropolitan centers that have passed through some threshold for self-sustaining growth, while recognizing that sectoral composition will shift with growth. He argued that the larger urban areas are more than proportionately places of creative entrepreneurship, from which he hazarded a broad hypothesis on the nature of regional growth patterns. The larger urban area, he said, is believed to invent, or at least innovate, to a more than proportionate degree and, therefore, to enjoy the rapid growth rate characteristic of the early stage of an industry's life cycle - one of exploitation of a new market. He said that "the economic base of the larger metropolis is . . . the creativity of its universities and research parks, the sophistication of its engineerin firms and financial institutions, the persuasiveness of its public relations and advertising agencies, the flexibility of its transportation networks and utility systems, and all the dimensions of 'infrastructure' that facilitate the quick and orderly transfer from old dying bases to new growing ones. Thus, as industry matures into a replacement market, the rate of job formation in that industry slows nationally and the local rate of job formation may slow even more if the maturing industry begins to decentralize - a likely development, especially in non-unionized industries, because with maturity the production process becomes rationalized and often routine. The high wage rates of the innovating area, quite consonant with the high skills needed in the beginning stages of the learning process, become excessive when the skill requirements decline and the industry, or parts of it, 'filters down' to the smaller, less industrially sophisticated areas where the cheaper labor is now up to the lesser occupational demands.

A filter-down theory of industrial location would go far toward explaining the small-towns' lament that they always get the slow-growing industries. The smaller, less industrially advanced area struggles to achieve an average rate of growth out of enlarging shares of slow-growth industries, originating as a by-product of the area's low wage rate attraction.

The larger, more sophisticated urban economies can continue to earn high wage rates only by continually performing the more difficult work. Consequently, they must always be prepared to pick up new work in the early stages of the learning curve - inventing, innovating, rationalizing, and then spinning off the work when it becomes routine. In its early stages an industry also generates high local incomes by establishing an early lead on competition, and thus smaller towns receiving "filtered" industries are, almost by definition, destined to have lower income levels than these prevailing in metropolitan areas.

In support of his formulation, and to illustrate the welfare consequences, Thompson reported on a variety of cross-sectional regression analyses that show that median family income in American cities is a positive function of locational educational levels, degree of manufacturing specialization, city size, the male labor force participation rate, and percentage of the population foreign born. Thus, larger cities have higher educational levels and more diversified occupational skills and general cultural environments, while this combines with the higher skill-mix in manufacturing, more capital per worker and greater productivity, kept high at the margin by union control of the labor supply, and by the combined product and factor price power of oligopolies and unions. Such a nexus of oligopoly, unions and capital-intensive production, particularly in heavy industry, contributes to the higher income levels of the big-city resident.

Continuing, and consistent with the theory, percent change in median family incomes, 1950-1960, was greatest in metropolitan areas approaching and surpassing the threshold for self-sustaining growth. High rates of male net immigration and a tight labor market also characterized such areas, consistent with Muth's formulation.

Some broad evidence for the city-size orientations of industry predictable if a hierarchical filtering process is operating can be gleaned by putting together materials from recent studies by the Stanford Research Institute (1968), Lichtenberg (1960) and Weiss (1969). Other data of a confirming kind are available in Stanback and Knight (1970). SRI computed location quotients by city-size clas for 4-digit SIC categories of industry using the data assembled by the U.S. Department of Commerce (1959) from the 1954 Census of Manufactures. From the relative sizes of these location quotients, a broad classification of industries by city-size orientation is possible, and is reported in Berry and Horton (1970). If the industries so classified are subdivided using Lichtenberg's categories of "dominant location factors," as detailed in an appendix to his study of industry in the New York metropolitan region, and Weiss's computation of the number of markets for the products of the industry within the U.S. (using the data in the Commodity Transport Survey of the 1963 Census of Transportation) is added, some most instructive results emerge. Industries concentrated in large metropolitan areas serve, as expected, national markets, deriving their locational advantage from externalities (economies of urbanization, localization, and industrial complexes) and from national market access. At the other extreme, small town-industry is of two kinds: resource-oriented processing (agriculture, forests) located so as to ensure maximum weight reduction before shipment of the product to national markets; and presumably "filtered" activities finding their locational advantage in low labor costs. In between, medium-sized cities appear to have to have concentrations of larger-scale agricultural processors, and a broad range of industries engaged in fabricating intermediates and producers' goods from metals.

How can a hierarchical filtering process be modelled? If, consistent with Tideman's argument outlined earlier, we are willing to postulate (and this does not seem unreasonable) that local growth based upon "filtering" industry comprises a combination of profit (=potential) maximizers seeking maximum access to national markets, but preempted from superior locations by early-comers, and industry filtering down the urban hierarchy in search of the lower labor cost characteristic of small towns, then the probability that a town i will recieve a filtering industry from town j can be written gravitationally as

$$\rho_{ij} = \beta \frac{P_i P_j}{D_{ij}^x} \tag{18}$$

From (18), it follows that the diffusion potential of town i, ρ_i is

$$\rho_i = \beta P_i \sum_j \frac{P_j}{D_{ij}^x} \tag{19}$$

In other words, the diffusion potential of town i is a joint function of its position in the urban hierarchy P_i and market potentials $\sum_j \frac{P_j}{D_{ij}^x}$.

Such a formulation can, of course, be rewritten in a variety of ways. For example, weights w_i and w_j can be attached to P_i and P_j or to some other measures of local "mass" M_i and M_j to index differences in the relative attractiveness or unattractiveness of origins and destinations. Also, alternative distance formulations may be preferred, for example the negative exponential $e^{-bD_{ij}}$ which appears in the entropy-maximizing form of the gravity model, or even the hierarchically-variable distance-decay function $e^{-b_j D_{ij}}$ where $b_j = f(P_j)$.

Having (19) one research strategy that is suggested is retrospective and static; industrial patterns can be compared with hierarchical position and potentials, or industrial change can be related to these same structural variables. It may, however, be more useful to consider the dynamics involved and postulate that the regional differences in rates of industrial immigration are related to changing diffusion potentials.

Then let

$$d\rho_i = \frac{\partial \rho_i}{\partial P_i} dP_i + \Sigma \frac{\partial \rho_i}{\partial D_{ij}} dD_{ij} \qquad (20)$$

Now

$$\frac{\partial \rho_i}{\partial D_{ij}} = - \beta P_i \ x \frac{P_j}{D_{ij}^{x+1}} \qquad (21)$$

So

$$d\rho_i = \frac{\rho_i}{P_i} dP_i - x\beta P_i \ \Sigma \frac{P_j}{D_{ij}^{x+1}} \ dD_{ij} \qquad (21)$$

$$= \rho_i \frac{dP_i}{P_i} \quad x\rho_i \frac{\Sigma \frac{P_j}{D_{ij}^x} \frac{dD_{ij}}{D_{ij}}}{\Sigma \frac{P_j}{D_{ij}^x}} \qquad (22)$$

If one postulates that the transportation system is improved homogeneously all $\frac{dD_{ij}}{D_{ij}}$ are equal. Call this increase λ

Then

$$d\rho_i = \rho_i \left(\frac{dP_i}{P_i} - x\lambda \right) \qquad (23)$$

Put

$$\frac{dP_i}{P_i} = \pi_i \qquad (24)$$

and

$$\frac{d\rho_i}{\rho_i} = \acute{\rho}_i \qquad (25)$$

Then

$$\acute{\rho}_i = \pi_i - x\lambda \qquad (26)$$

Thus, the relative improvement in potentials is equal to the difference between population growth and distance change multiplied by x, and the difference in <u>relative</u> potentials between town i and j is

$$\acute{\rho}_i - \acute{\rho}_j = \pi_i - \pi_j \qquad (27)$$

so that the relative improvement depends on the relative growth of population (provided, of course, the distance exponent x is independent of P).

This is clearly the simplest case, in which it would follow that the extent to which differences in manufacturing growth rates deviate from differences in rates of change of potentials indicate differences in relative accessibility change. Of course, it is not necessary to assume that all $\frac{dD_{ij}}{D_{ij}}$ are equal. Actual shifts in the operational measure of D_{ij} (cost, time, distance) can be measured and incorporated in the formulation and, indeed, empirical estimates of x not independent of P can also be incorporated. Then differences between city-to-city or region-to-region variations in manufacturing growth rates and changes in potential computed using (22) should reflect differences in communication paths and flows, and barriers to effective communication.

Industrial Decentralization and the Restructuring of Metropolitan America

The third component of industrial change to be explored is decentralization -- the changing relative location of industry within the city-suburb structure of metropolitan regions. This part of the exploration cannot be divorced from an understanding of urbanization processes, for we are confronted with classical models that appear to be substantially in need of revision.

The conventional wisdom about the nature of urbanization processes that has been widely accepted throughout the social sciences as a guide for understanding the logic of metropolitan growth and change was expressed no more straightforwardly than by Hope Tisdale when, in 1942, she wrote that

> urbanization is a process of population concentration. It proceeds in two ways: the multiplication of the points of concentration and the increasing in size of the individual concentrations . . . Just as long as cities grow in size or multiply in number, urbanization is taking place . . . Urbanization is a process of becoming. It implies a movement . . . from a state of less concentration to a state of more concentration.

From this widely accepted view, Louis Wirth derived a whole theory of the human consequences of urbanization, using as primary causal variables the size, density, and heterogeneity of the large city. And, consistent with the conventional wisdom, geographers and land economists in the interwar years produced a classic image of the North American city.

Much of this image involved models of commercial and industrial location within the city. Every city was was thought to require at its heart a strong, viable, growing central business district, with adjacent central industrial and commercial zones. At strategic locations within the main sectors of the city, and in a hierarchy of successively smaller communities and neighborhoods, there ought, the model said, to be a hierarchy of business districts providing for the shopping needs and service requirements of the population. Major traffic arteries were seen to support ribbons of highway-oriented businesses, heavy commercial uses, and specialized functional areas such as the "automobile row." Heavy industry was found next to major transportation facilities--ports, and railroad lines and spurs. Light industry was seen as gravitating to the suburbs.

While a variety of factors were said to be operating in creating this spatial structure (the locational advantages of ports, strategic positions on transportation arteries, availability of fuels, and raw materials all figure prominently in the writings of the period) the principal force was said to be that of urbanization economies.

Urbanization economies involve externalities: the consequences of the close association of many kinds of activity in cities, producing mutual scale advantages that lower the costs and improve the competitive position of individual firms. Such cost reductions were demonstrated, in a variety of studies, to arise from several sources. The argument is as follows:

> (1) Transport costs. Large cities have superior transport facilities offering significantly lower transport costs to regional and national markets. A city located at a focal point on transport networks is especially suited for easy assembly of raw materials and for ready distribution of products. There are thus market advantages for speedy and cheap distribution which may be accentuated by the advantages arising out of the size of the local consuming market. In the biggest cities, this local market will account for a very substantial share of service and residentiary activities and of the regionalized activities that need to minimize transport costs, since the local population may form a large part of the total national market. Greater London, for example has one-fifth of the British consumer market, and the New York metropolitan area has been called "one-tenth of a nation." In both these cases, these markets account for a significantly larger proportion of the higher quality and fashion demands of the respective nations than this. Thus it may be that, because of its market dominance, the larger city is the ideal place in which to locate a plant even if it is serving a national market, because substantial demands are local and the balance of the output is more easily distributed. A New York location, for example, will obviate the necessity of providing extra facilities in New York, such as showrooms and warehouses, that would otherwise be needed if the firm were located outside New York.

(2) Labor costs. The large-city labor market is extensive, diverse, and dynamic. in this circumstance, the labor demands of single firms are only a small part of the total demands for labor. Thus, recruitment is easier. This becomes especially important where firms have seasonally varying labor needs. The big-city labor market also offers a whole range of skills; consequently, the advantages of the city over the small town are obvious, even though the wage rates of the large city may be somewhat higher than those of the small town. Facilities for labor that are readily available in the large city may often also have to be provided in the small town, which raises the real labor costs to the firm.

(3) Advantages of scale. The larger the city, the higher the scale of services it can supply. Thus, fire and police, gas, electricity, and water, waste disposal, education, housing, and roads are all in general better in the larger city than in the small town. Firms developed in otherwise unindustrialized areas generally have to spend much more upon social capital than those that have been developed in large cities. These kinds of scale advantages came into play particularly in the industrialization of the largest metropolitan areas in the period following World War I, although interrupted by World War II. Urbanization economies are those that, in particular, have a magnetic effect upon those kinds of industries in which coal was replaced as motive power by electricity, and in which rail and water were replaced as a means of transportation by trucking over the roads. This new industry is "light" industry, concerned with the manufacturing consumers' goods. These goods are often branded and serve sheltered markets because of imperfect competition deliberately created by advertising mechanisms. For these industries, transport costs are a low proportion of total costs, because the materials used tend to be semi-processed and because the products are compact and have high value in relation to weight. Labor requirements are often unskilled or semi-skilled and the firms therefore are more likely to be attracted by large cities, especially by the markets and the marketing facilities of these large cities. Thus, there is a "snowballing" which tends to

> be accumulative, for, as light industry is attracted to large cities, the magnetic power of large cities for this kind of industry is increased, creating conditions favorable to further growth. This kind of an effect has been called "circular and cumulative causation."

But if these kinds of beneficial externalities were seen to be the forces drawing economic activities into cities, negative externalities were invoked as a means of explaining the spatial distribution of these activities within the city. Traffic congestion, parking problems, and difficulties of loading and unloading were all seen to militate against the most crowded parts of the urban region. Increased competition for labor, combined with the higher costs of urban living (especially rent and transportation) were seen as factors driving up money wages and impairing the competitive capabilities of activities with high wage bills. And, most significantly, the high land values resulting from intense competition for centrally located space in the growing city were seen to be instrumental in "sorting out" activities in the urban area, both within and beyond the central business district.

Industries remaining in or close to the central business district (CBD) were observed to be those which find a great advantage in the accessibility of the CBD, the point of convergence for all city and regional transport routes, offering unusual convenience for buyers and for the assembly of workers. But because industry in central business districts was seen to face severe competition for sites, a hierarchy of land usage was said to result. The "highest" uses in the hierarchy were seen to be those which derive the greatest advantage from centrality and which could therefore pay most by turning the advantage into profit and rents. Housing was low on the list, except for high-class apartment houses. Most types of industry were also seen to be driven out becuase, for them, the advantages of centrality are relatively small, and the costs of production are lower on the periphery of the city. Many shopping functions likewise were seen to move out to lesser farmers districts central to local community and neighborhood markets.

Therefore, as the process of land use competition took place, the higher uses occupied the central area, and the only economic activities that remained localized within the boundaries of the centra business districts were of the following kinds:

(1) <u>The financial district</u> of banks, insurance companies, lending institutions, brokers, and the like, relying upon speedy personal contact within a closely intertwined set of activities and relationships.

(2) <u>Specialized retail functions</u>, including department stores that require large population support at the point of convergence of population, firms that offer high-quality and rare goods which satisfy their need for great population support only at the point of maximum convergence of the population of the urban region.

(3) <u>Social and professional functions</u>, including the headquarters and main office function.

These activities tended to squeeze out housing and industry from the most accessible points. The manufacturing and service industry surviving in the inner city thus had very special features and tended tc be found on the edges of the central area in the back streets becaus land values there were lower, in the classical view. For this kind of industry, central location was imperative; there was no alternative. The industry had to be in the inner city because of its marketing needs, which could not be separated from its manufacturing functions. A simple showroom in the central area would not do. Alsc many central manufacturing industries were observed not to have moved because the inner areas supported relatively immobile skilled labor forces. Some also were said to be localized there because of the extreme subdivision of processes within the cluster of interdependent firms. Such firms use ground very extensively; their space requirements in relation to their level of output are very low. In this sense, they can use relatively obsolete multistory buildings in a variety of kinds of adapted space. This kind of space is, of course, available at the lowest rents until it is cleared and rebuilt for some other, more intensive, higher-paying kind of use. Industry in central areas thus tends to move from building to building, being displaced as those buildings are cleared for reuse and revenue. The scale of operation is generally small. Two good examples are printing and clothing.

For job printing in particular, the marketing factor is important, since rush jobs must be completed in a very short space of time. Firms are generally located on the margins of the professional and financial districts. Newspapers also tend to be produced in central locations, usually in very large, modern buildings near to the zone of highest land values. Here, centrality is related to the receipt of news. There has to be close contact among the editorial staff, printers, and the reporters who gather the news, and speedy distribution to the market is, of course, also imperative. On the other hand, periodicals, books, and standardized and regular kinds of job printing are not found in cities, but tend to decentralize into the suburban areas and rural and small-town locations.

Central-area clothing and related industries involve those branches where fashion and style are important--for example, ladies' dresses, gowns, furs, handbags, and fashion jewelry. Styles involves rapid changes in design, which means that output for stock is impossible; large varieties and small quantities are essential. Thus, small scale of activity is appropriate. There is no division of processes or line methods of production, and only a few persons work on a single garment. The work is highly skilled. Thus, in central London there are more than 2,000 firms making women's clothing. This kind of industry has no special space requirements. It can use converted obsolescent buildings, as the space required is small, but, on the other hand, skilled labor is essential. The trade is thus highly localized, with interdependence arising out of the provision of accessories by specialized and separate firms--embroidery, belts, buttons, and so on. Reliance on common specialists arises out of the fashion nature of the trade. The crucial point binding the industry to central locations is marketing, the essentiality of immediate displays to buyers. The industry has to be close to a place where buyers can cluster.

Manufacturing and selling are close together because the scale of firms is so small that they cannot afford to run separate premises and because selling and manufacturing are concerns of separate firms. Subcontracting is common, and people to whom the work is subcontracted need to be located close together. On the other hand, other branches of the clothing industry have quite different spatial distributions. Standardized products made by line methods, where there is subdivision of processes and where there are no requirements for a skilled labor force have decentralized.

The printing and clothing industries highlight the economies of urbanization and localization that held industry to the central areas of cities, surrounding the central business district. In the classic models, as the CBD grew and land values escalated, other activities would decentralize in a regular sorting-out process. Land values were considered the most important centrifugal force because they imposed increased costs upon industry. Moreover, in inner-city areas they resulted in cramped and restricted sites, in multistory buildings, in a shortage of space, in parking difficulties, in deficiency of light and air, and environmental pollution. Therefore, the pressure of space pushed out industry seeking space, light, and air to the periphery of the big city, to smaller towns, or to nonurban areas.

Steady decentralization has been observed in North American cities throughout the twentieth century, apparently attesting to the validity of the classical models. See, for example, the evidence in Figure 1.2, prepared using a table appearing in the recent work of urban economist Edwin S. Mills (1972). In this diagram, the average exponential density gradient of population and several types of employment for a sample of cities is plotted against time on semi-logarithmic graph paper, to show whether decentralization was continuous, or evidenced marked changes over the year. For each item plotted (except manufacturing during the Great Depression), continuous decentralization is seen to have been taking place as cities have grown and transportation has improved.

FIGURE 1.2

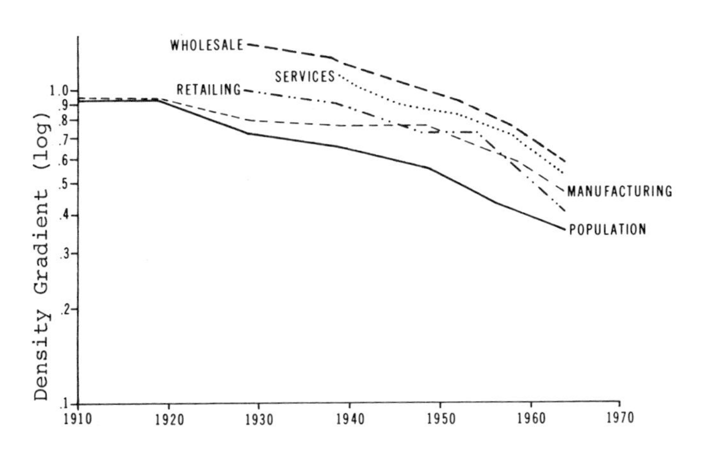

Findings such as these lead some commentators--for example, Banfield (1968)--to argue that all that is happening to cities, even today, is very logical, a straightforward extension of past trends to which the classic models still apply:

> Much of what has happened--as well as of what is happening--in the typical city or metropolitan area can be understood in terms of three imperatives. The first is demographic: if the population of a city

> increases, the city must expand in one direction or another--up, down, or from the center outward. The second is technological: if it is feasible to transport large numbers of people outward (by train, bus, and automobile) but not upward or downward (by elevator), the city must expand outward. The third is economic: if the distribution of wealth and income is such that some can afford new housing and the time and money to commute considerable distances to work while others cannot, the expanding periphery of the city must be occupied by the first group (the "well-off") while the older, inner parts of the city, where most of the jobs are, must be occupied by the second group (the "not well-off"). The word "imperatives" is used to emphasize the inexorable, constraining character of the three factors that together comprise the logic of metropolitan growth.

One indeed might conclude from inspection of Figure 1.2 that Banfield's argument is true--that since the last burst of inventions producing the upward and downward growth of the city (1880, the skyscraper; 1886, the subway; 1889, the elevator--see Eberhard, 1966), an outward urge had prevailed, with the last major contributing change the effects of the onset of the automobile after 1920.

But also notice that there were increases in the rate of decentralization for both manufacturing and retailing in the years following World War II. It is these cases which are the focus of much current discussion, indicating to some that radical restructuring of metropolitan forms may be taking place, and it is to this issue that we now turn. The point we want to discuss was raised perceptively by the historian Oscar Handlin (and Burchard, 1963), who argued that

> Some decades ago . . . a significant change appeared. The immediate local causes seemed to be the two wars, the depression, the new shifts in technology and population. However, these may be but manifestations of some larger turning point in the history of the society of which the modern city is part. The differences between city and country have been attentuated almost to the vanishing point. The movement of people, goods and messages has become so rapid and has extended over such a long period as to create a new situation. To put it bluntly, the urbanization of the whole society may be in the process of destroying the distinctive role of the modern city.

Are we indeed seeing such a radical restructuring of society, of which the changes in industrial and retail location are simply symptomatic?

Certainly, many statistics appear to support this view. Between 1947 and 1958, central cities of the old industrial heartland lost jobs and their suburbs gained as follows: Chicago, -18.5 and +49.4%; Philadelphia, -10.4 and +16.4; Detroit, -42.9 and +41.5; Boston, -15.3 and +33.5; and Pittsburgh, -25.3 and +18.1. In the period 1958-1963, the central cities of SMA's exceeding 250,000 population lost 338,000 manufacturing jobs, while 433,000 jobs were added to the suburbs. Between 1958 and 1967, net manufacturing employment changes for selected cities are reported in Table 1.4.

TABLE 1.4

Percentage Increase in Manufacturing Employment 1958-1967 (in thousands) Selected Metropolitan Areas

AREA	CENTRAL CITY	SUBURBS
Washington, D.C.	8.5	141.8
Baltimore, Md.	-5.9	22.0
Boston, Mass.	-11.8	17.0
Newark, N.J.	-10.3	12.2
Paterson-Clifton-Passaic, N.J.	-1.3	33.8
Buffalo, N.Y.	-1.9	3.4
New York, N.Y.	-10.3	36.0
Rochester, N.Y.	18.3	55.9
Philadelphia, Pa.-N.J.	-11.6	30.0
Pittsburgh, Pa.	-13.8	3.7
Providence, R.I.	4.5	12.4
Chicago, Ill.	4.0	51.6
Indianapolis, Ind.	23.1	36.3
Detroit, Mich.	-1.8	47.6
Minneapolis-St. Paul, Minn.	8.9	146.2
Kansas City, Mo.-Kans.	0	68.3
St. Louis, Mo.-Ill.	-14.9	41.4
Cincinnati, Ohio-Ky.-Ind.	10.6	2.7
Cleveland, Ohio	-5.3	42.6
Columbus, Ohio	18.1	0
Dayton, Ohio	12.9	99.5
Milwaukee, Wisconsin	-6.3	55.6
Miami, Fla.	5.7	116.0
Tampa-St. Petersburg, Fla.	3.8	280.0
Atlanta, Ga.	8.9	86.0
Louisville, Ky.-Ind.	16.7	48.4
New Orleans, La.	12.3	29.2
Dallas, Tex.	41.9	131.0
Houston, Tex.	41.6	12.6
San Antonio, Tex.	28.5	62.5
Los Angeles-Long Beach, Cal.	10.3	23.0
San Diego, Cal.	-24.5	27.5
San Francisco-Oakland, Cal.	-23.5	29.3
San Bernardino-Riverside-Ontario, Cal.	57.3	60.0
Denver, Colo.	6.9	112.7
Portland, Ore.-Wash.	16.5	68.0
Seattle-Everett, Wash.	-25.7	244.7

Most of the older central cities in the nation's "heartland" continued to experience substantial decline. Where there was growth from sources other than annexation, it was in the "rimland" regions of California, Florida, and Texas, where the greater city areas kept much of the "action" within the city limits.

To the reader of daily press and of the news magazines, loss of jobs in manufacturing and the decline of the retail function of the central business district presages disaster. Consider, for example, the following random clippings from stories appearing in 1972.

> the loss of the Giants was an almost comic sidelight to a more serious problem: the exodus of dozens of businesses, including many of the "Fortune 500," from New York to the surrounding suburbs or other cities. And since conditions in Gotham often forecast what may happen elsewhere, America's older cities are finding that they, too, are loosing [sic] some of their long-time corporate occupants to the burgeoning suburbs. The growing industrialization of the overwhelmingly white suburbs, coupled with the entrapment of minorities in the central cities, paints a picture with the word CRISIS splashed across it in boldface . . . If business [and suburban] communities continue in their present mindlessly selfish way, we will Los Angelesize our land, Balkanize our region's finances, and South Africanize our economy [Cassidy, 1972].

> The second great migration to the suburbs--the exodus of families and offices--has profoundly altered the economic life of cities, just as the great migration of middle-class families altered the social patterns. Four out of five new jobs created in major metropolitan areas in the past ten years have been outside the cities. This disproportion has been a major cause of the "urban crisis," damaging the ability of cities to pay their way and to provide employment for the people who live in them [Gooding, 1972].

> While concern has mounted in the last decade over the plight of the inner cities in vast metropolitan areas, the aging downtowns of hundreds of smaller cities have been quietly deteriorating with little national notice . . . losing businesses and office-users just as surely as their metropolitan cousins . . . "In the old days, the farmers all came to

> town on Saturday night, " . . . a long time resident recalled sadly. "Not any more, it's just dead." And a teenage girl driving what is locally known as the "the Circuit" said blithely: "This town is for nothing but old fogeys. It's a bummer" [Kneeland, 1972].

The press is also full of speculations about the reasons for the exodus of manufacturing and retailing, although there is far from any consistency in journalistic perceptions. Some see the shift from the pull of low land values to the push of urban problems:

> Business is moving out for a variety of reasons. The explanation immediately after World War II was the attraction of cheap land in the suburbs, permitting single-story factories that were convenient for truck loading. In recent years, the motive is more push than pull. Executives complain about the abominable phone service in many cities, horrendous commuting conditions, rapidly rising crime. Even bomb threats are mentioned, when GT&E actually had a bomb go off in the building, the bosses lost no time in making the final decision to get the hell out. Then there are problems with the work force. Many young women seem to be avoiding the big cities, while young execs no longer consider a move to the New York office a promotion; indeed, they demand differential pay to cover the increased cost of living. There is also the desire to get away from it all, which was one of the big reasons why Xerox moved its top men from Rochester to pastoral Connecticut: the company president felt that they would get a better perspective of the whole company from the new, more isolated locale. But the biggest appeal of the suburbs, of course, is that much of the population, housing and development is there, or headed there [Cassidy, 1972].

Others see the lure of new kinds of prestige locations, closer to the preferred places of residence of the businessmen and decision makers:

> Around the perimeter of O'Hare International Airport are any number of businesses that have O'Hare stuck in their name. The O'Hare area, which had no office space 10 years ago, now accounts for nearly 30 per cent of all suburban office space in the Chicago area. "Using O'Hare in the company name has the psychological aspect of giving the firm prestige," says Loren Trimble, an expert on the area's economic development. "O'Hare carries

the connotation of being near the world's busiest airport, of being modern and in the jet age." Proximity to the airport explains part of the phenomenal growth of the northwest suburbs in the last decade. O'Hare has had a very direct effect on the building of fancy hotels, convention facilities and office buildings in the area.

But it is only one of many reasons for the rapid industrial and commercial growth and, most experts agree, not the main one. Most maintain that things would be booming in the area without the world's busiest jetport.

"I say that the principal reason we have 500 new companies in Elk Grove Village is that it is close to where the boss lives," said Marshall Bennett, a partner in Bennett & Kahnweiler, industrial real estate brokers. His firm developed Centex Industrial Park in Elk Grove Village, the largest of its kind in the nation. The northwest suburbs are now Chicago's biggest competitor for the dollar of the industrialist, conventioner and nightclubber. The area is one of the four fastest growing areas in the country. Why this once rural economy grew so rapidly to an industrial and commercial giant is laid mostly to transportation. First came the people. They started the trek to the suburbs in the early 1950's. Later, the bosses and decision-makers in industry started bringing their factories closer to home. Old and dreary saw-toother-roof factories were abandoned in favor of horizontal layouts in spacious suburban buildings. They were mostly light industries that produced little noise, smoke, odor or other irritants to suburban residents. Many were warehouses. However, employes in these factories and warehouses have not been as keen as their bosses at being around O'Hare. Many of the workers, particularly the blue collar ones, still live in Chicago, while many of the northwest suburbanites still work in the city. This situation has led to considerable reverse commuting. For example, about 75 per cent of the 25,000 employes at Centex Industrial Park in Elk Grove commute from Chicago, while the majority of Elk Grove Village residents work in Chicago [Young, 1972].

Yet others look at identical situations and see only polarization centering on race:

A series of downtown events have highlighted the problem of the Chicago downtown area and raised fears that the ghostlike character of downtowns in most major U.S. cities will come to pass in Chicago as well. The abandoned

> appearance of downtowns throughout the country are a product of two factors: race and market and the interplay between them.
>
> The society that is in the majority judges an area's decline or renaissance in this country (whether it is a downtown area or a residential community) by the behavior of white residents and consumers--i.e., by an area's attraction and appeal to white people. This is attributable in part to our social and economic stratification and most particularly to the income distribution in our population.
>
> Therefore, since whites possess the wealth, the range of choices available to them, and the ways in which they act on such choices, tend to predestine an area's future. When whites decide to leave an area, a self-destructive process is perceived in white eyes to have been set in motion. This process feeds on itself and results in an inevitable reduction in the money necessary to support commercial facilities, pay for housing and building maintenance and make possible new construction. This in turn further accelerates the exodus of white populations and increases black populations, vacancies and abandonment [Meltzer, 1972].

What actually is happening and why? Before an answer can be provided, it must be said that all is not told by the statistics in the previous section. Indeed, there have been some dramatic changes of a directly contrary kind, and the counterpoint they play to the loss of manufacturing jobs and retail sales must be orchestrated into a balanced interpretation of contemporary urban dynamics. For example, who would acknowledge, reading the interpretations and data on Chicago already presented, that between 1953 and 1964 $750 millions were invested in construction and improvement of manufacturing plants in Chicago (Hartnett, 1971)?

The problem is that whenever a negative statistic is observed in the city today, someone hastens to cry wolf. Thus, when the usual annual summertime peak in new industrial investment commitments vanished in metropolitan Chicago in 1971 and 1972, and industrial vacancies in the central city started to mount, the Chicago Association of Commerce and Industry produced illustrations asserting that flight to the suburbs was the cause. This may be the source of major errors in interpretation, masking the fact that net change, plus or minus, is made up of many elements.

Consider the data on the sources of employment change in various parts of New York in the years 1967-1969 reported in Leone, (1972). Net change in each area is seen to represent the resolution of interacting birth, death, immigration, emigration, and local growth processes (Table 1.5). Clearly, given such complexity, any simple explanation is likely to be not only suspect, but almost undoubtedly faulty. Although Manhattan's CBD lost 10,625 jobs due to moveouts and 17,890 jobs due to businesses closing, 24,602 new jobs were created in new businesses and local nonmovers expanded their employment by 38,804! Much of this was associated with the office boom, to which we refer later. The decline in jobs in Brooklyn was due to an excess of deaths over births, because that area experienced net immigration. In Queens, too, there was net immigration, and the decline in that case was due to local nonmovers reducing their employment by 10,702.

TABLE 1.5

Employment Change in New York, 1967-1969

	Manhattan CBD	Brooklyn	Queens	New York SMSA
Births	+24,602	+5,882	+3,713	+45,146
Deaths	-17,890	-8,083	-3,901	-40,574
Immigrants	+5,851	+5,777	+6,892	+9,018
Emigrants	-10,625	-4,960	-3,078	-5,579
Local relocators	+8,136	+1,098	+961	+14,184
Nonmovers	+38,804	-1,338	-10,702	+11,696
Net change	+48,878	-1,624	-6,115	+33,891

SOURCE: Leone (1972), using data from the Dun & Bradstreet Corporation's "Dun's Market Identifier" files, prepared for analysis at the National Bureau of Economic Research.

The dynamic was different in each case. No wonder, then, the difficulty of many observers to square the gloom-and-doom arguments with a history that has transformed central business districts of the nation's major metropoli. Gross floor space in private office buildings in the Manhattan CBD increased from 126 million square feet in 1936 to only 128 million in 1950. But by 1960, the total was 160 million, by 1963, 184 (with an additional 18.8 in public buildings), and by 1970, 226 (with 20.8 in public buildings). In the Dallas CBD, comparable figures were 1936: 4.7 millions; 1950: 6.0; 1960: 15.6; 1963: 17.3; and 1970: 22.5 (Armstrong, 1972). In Chicago, 8.2 million square feet of office space was added to the Loop in the period 1967-1972. In the ten years 1960-1969, the valuation of new office space authorized for construction in the nation's major office centers was, in millions of dollars: New York SMSA, 1,659; Los Angeles SMSA, 1,220; Washington SMSA, 812; Chicago SMSA, 710; San Francisco SMSA, 601; Boston SMSA, 393; Detroit SMSA, 327; Atlanta SMSA, 279; Philadelphia SMSA, 250; Cleveland SMSA, 172; Seattle SMSA, 169; Milwaukee SMSA, 101. This investment boosted private and public office space by 44% in these SMSAs, which, along with Pittsburgh and St. Louis, account for over 70% of the headquarters and headquarters employment of the nation's top 500 industrials, and over 60% of all of the nation's central administrative office employment (another 7% is in Minneapolis-St. Paul, Houston, Dallas, Cincinnati and Kansas City; Armstrong, 1972).

These apparently contradictory trends, which evidently have defied the abilities of most popular commentators to comprehend their basic importance, are in fact a product of a revolutionary transformation of American metropolitan areas that renders conventional wisdom inoperable.

The initial development of road transport and highway improvements did indeed breed a continuous and gradual decentralization of people and jobs in the core-oriented metropolis through the first half of the twentieth century. But then came a series of technological changes that significantly modified the forces operating and produced the change in rates of decentralization observable in Figure 1.2.

On the manufacturing side, superhighways, large-scale trucks, and piggyback combination of road and rail transport reversed the pu of rail terminals and docks in central areas on industrial location. Previously, extremely efficient interregional transportation conspir with slow and costly local road transport to the industry to central terminal locations became as (or even more) accessible to the metropolitan region than was the traditional city center. With the remov of traditional constraints, the negative externalities of the centra city loomed large in choice, both perceptually and in fact. The massive decentralization of industry that followed was dictated more in its locational choices by social factors and prestige locations than by traditional dollars and cents. Low-status locations, poor, polluted, and black neighborhoods were left and avoided by new concerns

Again, the suggestion of the social dynamic is there replacing much of the economic dynamic of conventional wisdom. This social dynamic involves not only the avoidance of the central city by the majority white residents, but also the creative, style-setting leade ship. By leading rather than following, the new developers seek to establish the direction of metropolitan growth, to set the style and tone for the suburbs to follow, and to capitalize on the new opportunities thereby created.

While part of the social dynamic involves this creative develop ment of new opportunities on the expanding periphery, this is only possible because of the eagerness of most Americans to find a safe haven in the new forms of suburbia thereby created. As President Johnson's Commission on Crimes of Violence reported:

> If present trends are not positively redirected by creative new action, we can expect further social fragmentation of the urban environment, formation of excessively parochial communities, greater segregation of different racial groups and economic classes . . . and polarization of attitudes on a variety of issues. It is logical to expect the establishment of the 'defensive city,' the modern counterpart of the fortified

> medieval city, consisting of an economically declining central business district in the inner city protected by people shopping or working in buildings during daylight hours and 'sealed off' by police during nighttime hours. Highrise apartment buildings and residential 'compounds' will be fortified 'cells' for upper-, middle-, and high-income populations living at prime locations in the inner city, suburban neighborhoods, geographically removed from the central city, will be 'safe areas,' protected mainly by racial and economic homogeneity and by distance from population groups with the highest propensities to commit crime. Many parts of central cities will witness frequent and wide-spread crime, perhaps out of police control.

What all of this means, for example, is that the black resident of the metropolis finds himself in a central-city ghetto abandoned by both whites and, increasingly, employment. The flight of white city dwellers into the expanding peripheries of metropolitan regions is an accelerating phenomenon as minorities move toward majority status in the city center. The exurban fringes of many of the nation's urban regions have now pushed one hundred miles and more from the traditional city centers. More important, the core orientation implied in use of the term "central city" and "central business district" is fast on the wane. No longer is it necessary to have a single, viable, growing heart. Today's urban systems appear to be multinodal, multiconnected social systems in action, in which the traditional centralization of the population into metropolitan areas has been counterbalanced by a multifaceted reverse thrust of decentralization. The situation is very different from the period at the end of the nineteenth century from which we derive the concept of urbanization. Decentralization and an outward urge have replaced centralization and core orientation; differentiation and segregation substitute for the integrative role of the melting pot.

The essence of the new urban system is its linkages and interactions, as changed by changing modes of communication. Both places of residence and places of work are responding to social rather than to traditional economic dynamics. At the same time, new communications media, notably television, have contributed to cognitive change by providing the universal perception of decaying central cities, the new home of the former residents of the now-emptied periphery; the immediate on-the-spot experience of their riots; the careful documentation of their frustrations; and acute awareness of emerging separatist feelings. It is no accident that the suburbanization of white city dwellers has increased, supported by rising real incomes and increased leisure time. Similarly, decentralizing commerce and industry avoid the lower-status central-city location, and the new office complex seals itself in by defensive techniques. Gradients of distance-accretion are now beginning to replace those of core-centered distance-decay within the larger megalopolitan complexes as persons of greater wealth and leisure seek home and work among the more remote environments of hills, water, and forest, while most aspire to such settings as an ideal. In consequence, core-dominated concentration is on the wane; the multinode, multiconnection system is the rule, with the traditional multifunctional core simply a specialized one among many. It is the spontaneous creation of new communities, the flows that respond to new transportation arteries, the waves emanating from new industrial and retail growth centers, the mutually repulsive interactions of antagonistic social groups, the reverse commuting resulting as employment decentralizes, and a variety of other facets of social dynamics that today combine to constitute the new urban systems in America.

These trends and changes should, of course, be put back into the context of changing technology. Concentrated industrial metropoli only developed because proximity meant lower transportation and communication costs for those interdependent specialists who had to interact with each other frequently or intensively and could only do so on a face-to-face basis. But shortened distances also meant highe densities and costs of congestion, high rent, loss of privacy, and the like. As soon as technological change permitted, the metropolis was transformed to minimize these negative externalities. The decline of downtown retailing and of central-city manufacturing, alongside the office boom within the major metropolitan areas, are but manifestations of this fundamental transformation of American urbanism, for which a new definitive theory to match the conventional wisdom of the first half of the twentieth century has yet to be written.

CHAPTER 2

RELATIVE LOCATION IN THE AMERICAN SPACE-ECONOMY, 1950-1960

The fundamental hypothesis of this study, derived as it is from the literature reviewed in Chapter 1 is that industrial change in the American space-economy proceeds systematically along three dimensions of relative location: on a national scale, in terms of accessibility to national markets; on a regional scale with respect to the status of places in the urban hierarchy, and intra-regionally in terms of local access to the city center. The first dimension is that of the heartland-hinterland structure of the national economy; the second is that which, most fundamentally, gives order to the nation's major economic subregions; and the third is that which has most meaning in the analysis of the structure of the urban regions. The problem is to give these dimensions operational meaning, to analyze their interrelationships, to show how relative location changed during the period analyzed, and then to show how industrial change proceeded with respect to each dimension individually, to the set of dimensions simultaneously, and to those interacting components of relative location of demonstrable statistical significance (e.g. decentralization in medium-sized cities located within high access zones, etc.). We devote this chapter to the first set of these questions, involving the operational definition of the dimensions, their interrelationships, and the nature of changes in relative location in the United States in the period 1950-1960. Throughout, the units of observation are counties, for reasons of data availability and comparability.

ACCESSIBILITY TO NATIONAL MARKETS

To repeat what was said in Chapter I, there is good reason to expect industrial locations to vary with respect to accessibility to markets and resources. Any change in this accessibility should be an important source of locational change. In keeping with Harris's suggestions and Tideman's proofs, we chose to measure national access using the familiar population potential index, computed for each of the nation's 3,102 counties in both 1950 and 1960.

After much experimental cross-tabulation of the potential measures with indicators of industrial change, it also was determined that the statistically-significant relationships of industrial change to national accessibility could still be distinguished if, in the analysis, population potentials were grouped into three classes of access to national markets:

B_1 Low accessibility: Population potentials of less than 200,000 persons per mile.

B_2 Medium accessibility: Population potentials of 200,000-300,000 persons per mile.

B_3 High accessibility: Population potentials of more than 300,000 persons per mile.

This grouping also facilitated the subsequent handling and analysis of the complex total data system on which the study is based.

The following table shows the number and percentage of counties according to these three classes of population potential in 1950 and 1960:

	1950	1960	1960/1950
B_1	766 (24.6)	474 (15.3)	0.62
B_2	1,025 (33.0)	869 (28.0)	0.85
B_3	1,311 (42.4)	1,759 (56.7)	1.34
	3,102	3,102	1.00

The progressive increase in national market access during the decade is clear.

Change in Accessibility

Indeed, population potentials increased for every county between 1950 and 1960. After further experimentation, it was decided that statistical significance could be retained and change in potentials could be included in the analysis as a variable by simply distinguishing between a "growth" and a "rapid growth" category, viz:

b_1 Growth. Population potential increased by less than 20.0 per cent. 2,767 counties (89.2%).

b_2 Rapid Growth. Potentials increased by more than 20.0 per cent. 335 counties (10.8%).

The questions that then arose were how industrial changes related to initial patterns of national market access, whether the changes were sufficient to maintain the patterns of relationship as potentials changed, or whether the changes in potentials produced changes in the way in which industrial change unfolded.

STATUS WITHIN THE URBAN HIERARCHY

In Chapter I we also presented evidence on the hierarchical structuring of industrial locations within the nation's urban system as well as discussing mechanisms that would induce filtering of industries from one echelon of the hierarchy to another. The operational problem was to obtain the desired hierarchical stratification of counties.

This was accomplished in two stages. First, Berry's (1968) regionalization of the U.S. into Functional Economic Areas (FEA's) based upon commuting data collected in the 1960 census was used to provide a grouping of counties into urban-regional units. Second, the city serving as the economic core of the FEA was identified and its population was taken as an indicator of its relative hierarchical status. Justification for using population size as an index of hierarchical status will be found in the City Classification Handbook (Berry, 1972). These two steps enabled each county to have attached to it a measure of the hierarchical status of the urban center serving as the focus for the urban-regional unit (the FEA) within which the county was located, as well as a measure of the change in that urban center's population.

As in the case of national access, experimental cross-tabulation with the indicators of industrial change led to the conclusion that the statistically-significant relationships could be preserved by stratifying the size measure into three classes:

A_1 Small cities: FEA core city population less than 25,000.

A_2 Medium cities: FEA core city population 25,000-100,000.

A_3 Large cities: FEA core city population greater than 100,000.

The numbers and percentages of counties located within the FEA's of these size-classes of cities in 1950 and 1960 were:

	1950	1960	1960/1950
A_1	1,068 (34.4)	947 (30.5)	0.89
A_2	1,019 (32.8)	993 (32.0)	0.97
A_3	1,015 (32.7)	1,162 (37.5)	1.14
	3,102	3,102	1.00

The obvious immediate question was whether the relative decline of small centers and the growth of the largest ones had associated with it commensurate industrial changes.

Change in FEA Center Size

To facilitate tests of this question, four categories of changes in FEA central-city population size were distinguished, again after substantial statistical experimentation:

a_1 No change: 606 counties (19.5%).

a_2 Decline: FEA city population actually declined. 529 counties (19.1%).

a_3 Growth: FEA city population growth of 20 per cent or less. 8% counties (28.1%).

a_4 Rapid growth: FEA city population growth exceeded 20 per cent. 1,034 counties (33.3%).

A question that was then explored was whether industrial changes differed in counties located in urban regions whose central places experienced these different dimensions of population change in the decade.

SITUATION WITHIN URBAN FIELDS

Finally, a dimension along which centralization-decentralization trends could be assessed was provided by another outcome of the 1960 commuting studies: the percentage of the labor force living in each county who commuted to work in the central city of the FEA. The significance of this measure, as a key component of the nation's "regional welfare syndrome," has been documented elsewhere (Berry, 1973) and needs no further elaboration.

After experimental analysis, it was decided to stratify the commuting ties of each county to its FEA center into four levels of intensity:

C_1 No commuting: The county is beyond the daily commuting zone of the FEA center, but lies within its retail and/or wholesale market area.

C_2 Low commuting: Less than five per cent of the local labor force commutes to the FEA center.

C_3 Medium commuting: Between 5 and 20 per cent of the labor force commutes.

C_4 High commuting: More than 20 per cent commutes.

Unfortunately, commuting data were not available in 1950, so that no commuting change variable could be developed. The number and percentage of counties falling into the four commuting classes in 1960 was:

	1960	
C_1	657	(21.2)
C_2	1,283	(41.4)
C_3	647	(20.9)
C_4	515	(16.6)
	3,102	

Because of the absence of commuting change data, we had to use the criterion essentially as a spatial constant that enabled us to describe industrial change within urban fields. Thus, if decentralization of industry took place within urban fields, it would be revealed by different rates of industrial growth at the different levels of commuting intensity.

THE LOCATION DIMENSIONS CROSS-CLASSIFIED

The complete framework for the spatial analysis of industrial change is provided by looking at the three components simultaneously. Thus, the relative location of a county in the national space economy is viewed as being specified by its position in a three-dimensional matrix, and change in a county's relative location is viewed as a shift from one cell of the matrix to another. The advantage of such framework, from the geographer's point of view, is that it provides a matrix of locational categories that exhausts the national geographic area and that is free from ties to specific points on the map. Two counties of similar relative location can be close to each other or a thousand miles apart. Such a locational matrix should enable us to discern national trends and their variations. It also should help clarify significant directions of industrial change regardless of the particular location within the nation. In particular, it should enable industrial change to be analyzed with respect to the "initial" framework of relative location (1950), the "final" relativ locations of the counties (1960), and the changes in relative locati between 1950 and 1960. The questions to be asked are the following:

1. Is industrial change in the U.S. associated with relative location at a given point in time?
2. Is change in relative location associated with change in the industrial change variables?
3. Is industrial change associated simultaneously with both the initial locational advantage and the change in such advantage? Answers to these questions will only be understood by first understanding the combined categories of relative location and of locational changes in the three-dimensional matrix. Such combinations can first be examined pairwise, and then in their full three-way associations.

Accessibility and Hierarchical Status

The relationships between population potentials and FEA city size in 1950 and 1960 are as follows:

Population Potentials 1950

FEA City Size 1950	B_1 (Low)	B_2 (Medium)	B_3 (High)	Total
A_1 (Low)	491 (45.97)	351 (32.86)	226 (21.16)	1068
A_2 (Medium)	169 (15.10)	368 (32.89)	482 (43.07)	1119
A_3 (High)	106 (10.44)	306 (30.15)	603 (59.41)	1015
Total	766 (24.69)	1025 (33.04)	1311 (42.26)	3102

Population Potentials 1960

FEA City Size 1960	B_1 (Low)	B_2 (Medium)	B_3 (High)	Total
A_1 (Low)	201 (30.73)	338 (35.69)	318 (33.58)	947
A_2 (Medium)	112 (11.28)	241 (24.27)	640 (64.45)	993
A_3 (High)	71 (7.15)	290 (29.20)	801 (80.66)	1162
Total	474 (15.28)	869 (28.01)	1759 (56.70)	3102

Apparently, the overall expansion of the zone of highest market potential has been accompanied by large-city growth, as can be discerned by examining the shifts in the percentage of counties in each of the two-way classes in 1950 and 1960:

			1950	1960	1960/1950
A_1B_1	Small F.E.A.	Low Potential	15.83%	9.38%	.59
A_1B_2		Medium	11.32	10.90	.96
A_1B_3		High	7.29	10.25	1.41
A_2B_1	Medium	Low	5.45	3.61	.66
A_2B_2		Medium	11.86	7.77	.66
A_2B_3		High	15.54	20.63	1.33
A_3B_1	High	Low	3.42	2.29	.67
A_3B_2		Medium	9.86	9.35	.95
A_3B_3		High	19.44	25.82	1.33

A cross-classification of changes in potentials and changes in FEA center populations provides an extra nuance, however:

	FEA Size Change			
Poten. Change	a_1 (No change)	a_2 (Decline)	a_3 (Growth)	a_4 (Rapid Growth)
b_1 (Growth)	491 (81.0)	569 (96.1)	815 (93.7)	892 (86.3)
b_2 (Rap. Growth)	115 (19.0)	23 (3.9)	55 (6.3)	142 (13.7)

Clearly there were significant shifts involving rapid city growth in areas with rapidly increasing potentials. Did this in turn, produce, rapid industrial growth of those areas? And were industrial dynamics in these areas different from changes taking place in these counties at the other extremes--e.g. the 115 counties with rapidly-growing potentials, but stagnant FEA cities; or the 23 with rapid growth of potentials but actual declines in FEA size? Again, these are questions to be addressed later.

Hierarchical Status and Local Access

Comparable two-way tables classifying counties by FEA size and by commuting intensities are shown on page 54, following.

		1960 Commuting			
FEA City Size 1950	C_1 (No Commuting)	C_2 (<5%)	C_3 (5-20%)	C_4 (>20%)	Total
A_1 (Small)	605 (56.65)	226 (21.16)	129 (12.08)	108 (10.11)	1068
A_2 (Medium)	17 (1.67)	508 (49.85)	276 (27.08)	218 (21.39)	1019
A_3 (High)	35 (3.45)	549 (54.09)	242 (23.84)	189 (18.62)	1015
Total	657 (21.18)	1283 (41.36)	647 (20.86)	515 (16.60)	3102

FEA City Size 1960	C_1 (No Commuting)	C_2 (<5%)	C_3 (5-20%)	C_4 (>20%)	Total
A_1 (Small)	603 (63.67)	169 (17.85)	98 (10.35)	77 (8.13)	947
A_2 (Medium)	18 (1.81)	488 (49.14)	269 (27.09)	218 (21.95)	993
A_3 (High)	36 (3.10)	626 (53.87)	280 (24.10)	220 (18.93)	1162
Total	657 (21.18)	1283 (41.36)	647 (20.86)	515	3102

That the greatest shifts involve FEA city growth in high-commuting areas is borne out by the two-way listing of relative changes:

			1950	1960	1960/1950
A_1C_1	Small F.E.A.	No Commuting	19.50%	19.44	1.00
A_1C_2		<5% Commuting	7.29	5.45	.75
A_1C_3		5-20% Commuting	4.16	3.16	.76
A_1C_4		>20% Commuting	3.48	2.48	.71
A_2C_1	Medium F.E.A.	No Commuting	.55	.58	1.05
A_2C_2		<5% Commuting	16.38	15.73	.96
A_2C_3		5-20% Commuting	8.90	8.67	.97
A_2C_4		>20% Commuting	7.03	7.03	1.00
A_3C_1	Large F.E.A.	No Commuting	1.13	1.16	1.03
A_3C_2		<5% Commuting	17.70	20.18	1.14
A_3C_3		5-20% Commuting	7.80	9.03	1.16
A_3C_4		>20% Commuting	6.09	7.09	1.16

Stability of FEA city populations in those parts of the country where the retail and wholesale market areas are far more extensive than the daily commuting areas is evident. As the following table shows, 601 counties fall into the no-commuting: stable FEA city population category:

FEA Size Change

	Commuting in 1966	a_1	(No Change)	a_2	(Decline)	a_3	(Growth)	a_4	(Rapid Growth)	Total
C_1	(No Com'g)	601	(91.48)	21	(3.20)	16	(2.44)	19	(2.89	657
C_2	(0-5)	3	(.23)	294	(22.92)	411	(32.03)	575	(44.82)	1283
C_3	(5-20)	1	(.15)	148	(22.87)	249	(38.48)	249	(38.48)	647
C_4	(20)	1	(.19)	129	(25.05)	194	(37.67)	191	(37.09)	515
	Total	606	(19.54)	592	(19.08)	870	(28.05)	1034	(33.33)	3102

On the other hand, there are approximately twice as many counties contained in the daily labor markets of the rapid-growth FEA cities as there are in the FEA's whose populations are declining. What are the differences in industrial growth among these several cases? This, again, is a question to be addressed.

National and Local Access

The final two-way classification is that of national and local access:

Commuting Zones	Low (Population Potential 1950)	Medium (Population Potential 1950)	High (Population Potential 1950)	Total
C_1 (No Commuting)	335 (43.73)	179 (17.46)	143 (10.91)	657
C_2 (<5%)	250 (32.64)	481 (46.93)	552 (42.10)	1283
C_3 (5-20%)	84 (10.97)	222 (21.66)	341 (26.01)	647
C_4 (>20%)	97 (12.66)	143 (13.95)	275 (20.98)	515
Total	766 (100.00)	1025 (100.00)	1311 (100.00)	3102

Commuting Zones	Low (Population Potential 1960)	Medium (Population Potential 1960)	High (Population Potential 1960)	Total
C_1 (No Commuting)	234 (49.37)	198 (22.78)	225 (12.79)	657
C_2 (<5%)	141 (29.75)	403 (46.38)	739 (42.01)	1283
C_3 (5-20%)	47 (9.92)	148 (17.03)	452 (25.70)	647
C_4 (>20%)	52 (10.97)	120 (13.81)	343 (19.50)	515
Total	474	869	1759	3102

Again, the overall influence is that of increasing population potentials, with the greatest increases in the high commuting areas, as both the two-way and the growth cross-tabulations reveal:

		1950	1960	1960/1950
B_1C_1	Low Potential No Commuting	10.80	7.54	.70
B_1C_2	5% Commuting	8.06	4.55	.56
B_1C_3	5-20% Commuting	2.71	1.52	.56
B_1C_4	20% Commuting	3.13	1.68	.54
B_2C_1	Medium Potential No Commuting	5.77	6.38	1.11
B_2C_2	5% Commuting	15.51	12.99	.84
B_2C_3	5-20% Commuting	7.16	4.77	.67
B_2C_4	20% Commuting	4.61	3.87	.84
B_3C_1	High Potential No Commuting	4.61	7.25	1.57
B_3C_2	5% Commuting	17.80	23.82	1.34
B_3C_3	5-20% Commuting	10.99	14.57	1.33
B_3C_4	20% Commuting	8.86	11.06	1.25

Commuting Zones	Potential Change b_1 (Growth)	b_2 (Rapid Growth)	Total
C_1 (No Commuting)	535 (19.34)	122 (36.42)	657 (21.18)
C_2 (5%)	1198 (43.30)	85 (25.37)	1283 (41.36)
C_3 (5-20%)	603 (21.79)	44 (13.13)	647 (20.86)
C_4 (20%)	431 (15.58)	84 (25.07)	515 (16.60)
Total	2767	335	3102

Is industry decentralizing as the number of low-potential no-commuting counties is declining? Has the rapid growth of national market access in the more centrally-located counties produced further industrial growth in these locations? What is the balance of trends? What is the relative influence of national market, hierarchical and local access shifts? These are among the questions to be explored.

The Three-Dimensional Scheme

So far we have simply produced the different two-way cross-classifications of relative location in 1950 and 1960, and their shifts in the decade. The three-way breakdowns that follow show the locationally-selective character of relative growth and decline clearly enough:

	1950	1960		1950-1960
$A_1B_1C_1$	329	230	$a_1b_1C_1$	486
$A_1B_1C_2$	80	26	$a_1b_1C_2$	3
$A_1B_1C_3$	39	18	$a_1b_1C_3$	1
$A_1B_1C_4$	43	17	$a_1b_1C_4$	1
$A_1B_2C_1$	168	191	$a_1b_2C_1$	115
$A_1B_2C_2$	94	83	$a_1b_2C_2$	-
$A_1B_2C_3$	50	32	$a_1b_2C_3$	-
$A_1B_2C_4$	39	32	$a_1b_2C_4$	-
$A_1B_3C_1$	108	182	$a_2b_1C_1$	21
$A_1B_3C_2$	52	60	$a_2b_1C_2$	289
$A_1B_3C_3$	40	48	$a_2b_1C_3$	141
$A_1B_3C_4$	26	28	$a_2b_1C_4$	118
$A_2B_1C_1$	3	1	$a_2b_2C_1$	-
$A_2B_1C_2$	105	69	$a_2b_2C_2$	5

	1950	1960
$A_2B_1C_3$	24	19
$A_2B_1C_4$	37	23
$A_2B_2C_1$	5	4
$A_2B_2C_2$	206	134
$A_2B_2C_3$	93	48
$A_2B_2C_4$	64	55
$A_2B_3C_1$	9	13
$A_2B_3C_2$	197	285
$A_2B_3C_3$	159	202
$A_2B_3C_4$	117	140
$A_3B_1C_1$	3	3
$A_3B_1C_2$	65	46
$A_3B_1C_3$	21	10
$A_3B_1C_4$	17	12
$A_3B_2C_1$	6	3
$A_3B_2C_2$	181	186
$A_3B_2C_3$	79	68
$A_3B_2C_4$	40	33
$A_3B_3C_1$	26	30
$A_3B_3C_2$	303	394
$A_3B_3C_3$	142	202
$A_3B_3C_4$	132	175
Total	3102	3102

	1950-1960
$a_2b_2C_1$	7
$a_2b_2C_2$	11
$a_3b_1C_3$	15
$a_3b_1C_4$	394
$a_3b_1C_1$	230
$a_3b_1C_2$	176
$a_3b_2C_3$	1
$a_3b_2C_4$	17
$a_3b_2C_1$	19
$a_3b_2C_2$	18
$a_4b_1C_3$	13
$a_4b_1C_4$	512
$a_4b_1C_1$	231
$a_4b_1C_2$	136
$a_4b_2C_3$	6
$a_4b_2C_4$	63
$a_4b_2C_1$	18
$a_4b_2C_2$	55
Total	3102

CHAPTER 3

THE METHODOLOGY EXEMPLIFIED

The relative location of each county in the nation's economic space thus is viewed as three-dimensional, involving national market access, position within a hierarchical status system, and local access within the functional economic areas into which the national economy may be subdivided on the basis of daily labor market relationships. Each of these dimensions has been provided with an operational measurement. Counties may change their relative location in the nation's space-economy as the pattern of national market access shifts, as the rank of urban centers rises or falls, and as local access is changed. These shifts may have associated with them either further heartland growth or dispersion of industry to hinterland regions on a national scale. They may be associated with further concentration of activity in major metropolitan centers, or with diffusion of activity to lower echelons of the urban hierarchy. And they may be related to further centralization in urban centers, or to decentralization to the suburbs and beyond.

Such responses are to be understood in terms of the locational economies that each of the dimensions represents, and the changes in these economies that accompany the changing nature of the national economy. To review briefly, the hierarchical relationship is built into the analysis because of the presumption that counties benefit from the economies accruing to city size. The external economies present in cities, the availability of services in them, their indus trial mix, their employment opportunities, as well as many other eco nomic and social characteristics, have both direct and indirect infl ences on the economies of counties that are located in their urban fields. The benefits that counties can derive from being located in urban fields depend, in turn, on the size of the central place; the larger the size, the greater the expected benefits. They also depen on the location of the counties within the urban field; counties within the field will tend to specialize in those activities for which they have the highest comparative advantage relative both to that of the urban focus and to other counties within the field. Hie archical status and within-region specialization are not independent generally, the larger the urban core, the greater the number of alte natives for economic specialization, and the greater the possibility of diversification of the economic base. With increased growth in the urban focus some activities are forced out of it. It is likely that they will relocate either in more remote parts of the field, or in centers lower in the urban hierarchy. The result is similar when diseconomies of different kinds appear in the urban center.

These processes are enhanced by two major trends that have unfolded in all modern economies in the last three decades. The mecha nization of agriculture which gave momentum to the urbanization process is continuing and is further accompanied by organizational change. The result is an ever-increasing significance of economies of scale in agriculture. Counties in the rural hinterlands of urban foci are forced to make adjustments if they are to prevent loss of population. Such adjustments can be made only by transforming the economic base to include a greater proportion of secondary and tertiary activities. Such a transformation will be less difficult for counties that are accessible to large urban centers than for centers that lack a viable local economic focus.

The second trend is associated with the continued growth of cities since World War II, the rise of a service economy, and accelerated suburbanization. Suburbanization is more likely to affect counties in the hinterland of a large urban focus. Indications are that when suburbanization occurs in a small urban center, the major beneficiary in terms of economic growth is the county in which the urban center is located. If other counties are to be affected by suburbanization, the urban focus must be greater in size.

The effects of being associated with a given level of the urban hierarchy and of relative location within the nation are further affected by the relative accessibility of counties within the nation. The greater the relative accessibility of a county within the nation, the more it is likely to benefit from suburbanization, from filtering-down effects and from economies available at its own or other urban foci.

The economies associated with relative location are the result of the cumulative effects of technological innovations and changes in the organization of production. These economies are not in a steady-state; they are changing. The rise and fall, in absolute or in relative terms, of regions within the nation is proof enough of this. Such changes depend, in turn, on changes in the technology of both production and transportation that set in motion both centrifugal and centripetal forces simultaneously, the overall dimensions of growth in any period depending upon the particular balance of the two at that time.

One of the ways in which centralization and decentralization tendencies can be discerned is by examining growth differentials between areas preclassified as core and periphery. This was the research strategy of Creamer, discussed in Chapter I. Another is to attach to each county several measures of relative location and of changes in relative location, and to use these as the "independent" variables "explaining" manufacturing change. This is the research strategy adopted here.

THE DEPENDENT VARIABLES

To elicit the full richness of the patterns of manufacturing change, three sets of manufacturing variables--seven variables in all -- were selected as the "dependent" variables to be "explained." The first set considers employment, and includes two variables: change in the total number of manufacturing employees by place of residence between 1950 and 1960; and change in the total number of manufacturing employees by place of work between 1952 and 1962. Change of the first kind could indicate one of two things: a shift of the labor force from agricultural occupations to manufacturing ones, especially in rural areas, and, particularly in larger metropolitan areas, residential change of manufacturing employees.

The second set of variables deals with changes in the number of manufacturing establishments: the change in the total number of manufacturing establishments; change in the total number of "medium-siz establishments, with 20 to 99 employees; and change in the number of "large" plants, with more than 99 employees. These three measures should reveal different trends in manufacturing. Since plants of different employment sizes have differential locational requirements the influence of different relative locations on manufacturing chang can thus be revealed more completely than by taking any single indicator.

The last set of variables was chosen in order to weigh manufacturing change in terms other than employment or employment-size. This set includes two variables: the change in manufacturing payrolls; and the change in the value added by manufacturing. Again both changes are for the period 1952 to 1962. While the first two sets of variables enable us to analyze the influence of relative loc tion manufacturing change in terms of the end results of locational decisions by both entrepreneurs and workers, the last set provides economic weight to such decisions.

Table 3.1 shows the average change for each variable, its standard error, and a categorization of counties into classes of change. The categories of change in each variable were determined on a qualitative basis. Given the frequency distribution of each of the variables, it was decided initially to differentiate between three categories of change: 1. decline or no change; 2. growth; and 3. rapid growth. Later, the growth and rapid growth categories were combined for purposes of analysis, because the basic research strategy was to analyze the continuous relationships between dependent and independent variables in a regression sense, but to use a variance-covariance framework to elicit the details of the changes that were unfolding.

MODES OF ANALYSIS

Relative location was measured using two dimensions for 1950 and three for 1960 (A, hierarchical status; B, national access; and C, local access in 1960) while changes in location were measured for two dimensions (a, hierarchical change; b, change in national market access). Thus, using 1960 local commuting as a spatial "constant," three threeway crossed designs were developed as a framework for analysis of each of the seven manufacturing change variables:

1950	1960	1950 - 1960
A_1, A_2, A_3	A_1, A_2, A_3,	a_1, a_2, a_3, a_4
B_1, B_2, B_3	B_1, B_2, B_3	b_1, b_2
C_1, C_2, C_3, C_4	C_1, C_2, C_3, C_4	C_1, C_2, C_3, C_4
AXBXC=36 Groups	AXBXC=36 Groups	axbxc=32 Groups

In addition, a more complex design was obtained by cross classifying the initial relative location in 1950 with the change in the components of relative location between 1950 and 1960, resulting in 288 relative location groups: $(A \times a) \times (B \times b) \times C = (3 \times 4) \times (3 \times 2) \times 4 = 288$. Only the cross classification of relative locations for the 1950 beginning and the 1960 the end-point resulted in designs with-

TABLE 3.1

Industrial Change Variables: Average Change, Standard Error, Categories of Change

Variable	X̄	Standard Error	Decline or No Change	Growth	Rapid Growth	Total
			Number [and Percent] of Counties			
001	8.861	.300	1159 [37.36]	936 (<5.0)[2] [30.17]	1007 (>5.0) [32.46]	3102
002	1.534	.045	1274 [41.07]	1323 (<2.0) [42.65]	505 (>2.0) [16.28]	3102
003	1.538	.056	869 [28.01]	1902 (<2.0) [61.32]	331 (>2.0) [10.67]	3102
004	.945	.039	2001 [64.51]	770 (<1.6) [19.79]	331 (>1.6) [15.75]	3102
005	.691	.022	2317 [77.92]	368 (<1.0) [11.83]	317 (72.0) [10.25]	3102
006	2.173	.066	953 [30.72]	931 (<2.0) [30.01]	1218 (>2.0) [39.26]	3102
007	2.194	.073	1041 [33.56]	905 (<2.0) [29.18]	1156 (>2.0) [37.27]	3102

The variables are:

1: Change in the Total Number of Manufacturing Employees (place of residence).. 1960/1950
2: Change in the Total Number of Manufacturing Employees (place of work)....... 1962/1952
3: Change in the Total Number of Manufacturing Establishments................. 1962/1952
4: Change in the Total Number of Manufacturing Establishments with 20-99 employees 1962/1952
5: Change in the Total Number of Manufacturing Establishments with > 99 employees 1962/1952
6: Change in the Total Manufacturing Payroll 1962/1952
7: Change in the Total Manufacturing Value-added 1962/1952

[2]Numbers in parentheses are the percentage growth rates for the "growth" and "rapid growth" classes.

out "empty" groups. The third crossed design contained four empty cells, and in the fourth case many of the 288 logical groups were without members.

Recall that the basic intent was to determine whether manufacturing change was patterned similarly or differently with respect to relative location in 1950 and 1960, and whether changes in relative location had parallel shifts in types of manufacturing change, or whether there was apparently change in the nature of change. From such evidence, it was hoped to determine whether, in a period of economic growth such as the 1950's, a systematic spread of growth took place along the three relative location dimensions, involving therefore: (a) <u>diffusion</u> from high levels of the urban hierarchy to lower levels, (b) <u>dispersion</u> of growth from areas of high relative accessibility, and (c) <u>decentralization</u> from urban centers to areas further away in the urban field. A related question was whether, in a period of economic growth, the gap between core and peripheral areas in the nation's economic space narrowed.

At this juncture, a cautionary note is in order, however. The relative location/relative growth framework of this study does not actually observe the above processes unfolding step by step. What is shown is, for example, whether or not growth occurred at lower levels of the urban hierarchy. If this is the case, it is <u>inferred</u> that diffusion is taking place. The occurrence of dispersion can be inferred by observing growth at different levels of relative accessibility, and decentralization can be evaluated by looking at growth in different zones within urban fields. Furthermore, it is possible to examine the simultaneous interaction of diffusion and dispersion (that is to examine the pattern of change in a group of counties associated with a given level of the urban hierarchy but possessing different levels of relative accessibility and vice versa), of diffusion and decentralization, and of dispersion and decentralization.

It is also possible to examine the three-way interactions between diffusion, dispersion and decentralization.

This is accomplished by a series of cross-classificatory analyses in which, first, chi-square procedures are used to assess the nature of the changing relative locations of counties in the nation's economic space and, second, variance-covariance models are used to explore the changes in industrial growth rates across the classes of relative locations and locational changes.

AN EXAMPLE: CHANGE IN NUMBERS OF MANUFACTURING ESTABLISHMENTS

Change in the number of manufacturing establishments will be used as an example to illustrate these methods of analysis. Chapters 4-6 will then be devoted to full analysis of the remaining six dependent variables. First, the relationships between manufacturing change and each of the relative location dimensions are considered separately, with contingency tables and chi-square analysis the principal tools (Tables 3.2, 3.3 and 3.4). Second, relationships to pairs of locational dimensions and locational changes are considered simultaneously, again in a contingency and chi-square framework (Tables 3.5, 5.6 and 3.7). Third, analysis of variance is undertaken of the relative contributions of the location dimensions to differences in manufacturing growth rates among locational classes of counties (Table 3.8ff.). And finally, covariance models are used to obtain the best possible predictions of growth differentials.

Hierarchical Diffusion

During the 1950's, growth in numbers of industrial establishments was found at all levels of the urban hierarchy, although the proportion of counties growing varied from 79.4% of those within the FEA's of the highest-level centers to only 62.8% of those located in the FEA's of the lowest-level centers, as the uppermost contingency table as Table 3.2 indicates. These proportions differ significantly from

TABLE 3.2

Initial Location in the Urban Hierarchy and Change in the Total Number of Manufacturing Establishments 1952 - 1962.

Manufacturing Change	A_1		A_2		A_3		Total	
Decline	397	(37.17)	263	(25.81)	209	(20.59)	869	(28.01)
Growth	671	(62.83)	756	(74.19)	806	(79.41)	2233	(71.99)
Total	1068		1019		1015		3102	

	Expected Frequencies			Deviations From Observed Frequencies			Chi-Square Components		
	A_1	A_2	A_3	A_1	A_2	A_3	A_1	A_2	A_3
Decline	299	285	285	+98	-22	-76	32.12	1.70	20.27
Growth	769	734	730	-98	+22	+76	12.49	.66	7.91

X^2 (2)= 75.15 Sig.<.001

Location in the Urban Hierarchy 1960 and Change in the Total Number of Manufacturing Establishments 1952 - 1962.

Manufacturing Change	A_1		A_2		A_3		Total	
Decline	364	(38.44)	240	(24.17)	265	(22.81)	869	(28.01)
Growth	583	(61.56)	753	(75.83)	897	(77.19)	2233	(71.99)
Total	947		993		1162		3102	

	Expected Frequencies			Deviations From Observed Frequencies			Chi-Square Components		
	A_1	A_2	A_3	A_1	A_2	A_3	A_1	A_2	A_3
Decline	265	278	326	+99	-38	-61	36.98	5.19	11.41
Growth	682	715	836	-99	+39	+61	14.37	2.13	4.45

X^2 (2)= 74.53 Sig.<.001

Change in the Urban Hierarchy 1950 - 1960 and Change in the Total Number of Manufacturing Establishments 1952 - 1962.

Manufacturing Change	a_1 & a_2		a_3		a_4		Total	
Decline	359	(29.97)	218	(25.06)	292	(28.24)	869	(28.01)
Growth	839	(70.03)	652	(74.94)	742	(71.76)	2233	(71.99)
Total	1198		870		1034		3102	

	Expected Frequencies			Deviations From Observed Frequencies			Chi-Square Components		
	a_1 & a_2		a_3	a_1	a_2	a_3	a_1 & a_2	a_3	a_4
Decline	336	244	289	+23	-26	+3	1.57	2.77	.03
Growth	862	626	745	-23	+26	-3	.61	1.08	.01

X^2 (2)= 6.07 Sig.<.05

those that would have been expected if growth and hierarchical status were unrelated. If this had been the case, the growth proportions could have been calculated from the marginal frequencies of the table, using the usual formula for the joint probability of independent events. Such a calculation would have produced the "expected" frequencies in the second row of contingency tables in Table 3.2. The differences between observed and expected frequencies are highly significant, however (chi-square <0.001), with the deviations shown in the middle contingency table in the second row, and the contributions of the deviations to the total chi-square in the third table in the second row. The deviations reveal that counties located in the FEA's of high-status centers experienced more industrial growth than expected, while counties located in the FEA's of low-status centers declined more than expected. Industrial growth in the 1950's was positively related to location within the economic sphere of high-status centers at the beginning of the decade.

Almost identical results are obtained by comparing industrial growth during the 1950's with location within the FEA's of centers of different hierarchical status at the end of the decade in 1960 as the middle set of contingency tables in Table 3.2 reveal. The chi-square statistic and the deviations between observed and expected frequencies are almost identical, despite the fact that the number of counties located within the FEA's of lower-status centers (groups A_1 and A_2) declined by 147, while the number of counties located within top-echelon FEA's (group A_3) increased by that number. In short, while many counties were improving their relative location, the positive relationship between hierarchical status and industrial growth prospects remained the same, meaning that a growing number of counties were now within the status range most likely to benefit from industrial growth. Now recall that we began with the hypothesis that in a period of rapid economic growth, industrial change should diffuse hierarchically from larger centers to lower levels of the urban hierarchy. The results presented in Table 3.2 apparently belie this hypothesis. Rather than accelerated growth filtering down, counties "filtered up." Growth prospects did not increase for lower-status counties; instead, counties' growth prospects increased as the economic centers of their FEA's improved their hierarchical status.

The third set of comparisons in Table 3.2 is of industrial growth with the rate of population change of the economic centers of the FEA's in which the counties are located. There are proportionately more declining manufacturing counties located in FEA's with declining or stagnant economic cores (a_1 and a_2), and conversely, there is more industrial growth in the growing FEA's (a_3 and a_4) than would be expected on the basis of the marginal frequencies in the contingency tables. This confirms the previous conclusions. Counties located within the FEA's of urban centers improving their hierarchical status were far more likely to experience industrial growth than counties within the FEA's of stagnant or declining centers.

Dispersion

Table 3.3 shows similar things to Table 3.2. Relative location levels of all counties were improved substantially during the 1950's, and relative accessibility was clearly associated with growth in the total numbers of manufacturing establishments. The greater the relative accessibility of counties in 1950, the greater the proportion that grew between 1950 and 1960 -- more than expected on the basis of the joint probability of independent events computed from the marginal frequencies -- and conversely, the less the accessibility in 1950 the greater the likelihood of decline. By 1960, the results were even clearer. A far greater proportion of those counties that had grown in the decade were now found in high potential areas.

TABLE 3.3

Initial Relative Accessibility and Change in the Total Number of Manufacturing Establishments 1952 - 1962.

Manufacturing Change	B_1		B_2		B_3		Total	
Decline	297	(38.77)	304	(29.66)	268	(20.44)	869	(28.01)
Growth	469	(61.23)	721	(70.34)	1043	(79.56)	2233	(71.99)
Total	766		1025		1311		3102	

	Expected Frequencies			Deviations From Observed Frequencies			Chi-Square Components		
	B_1	B_2	B_3	B_1	B_2	B_3	B_1	B_2	B_3
Decline	215	287	367	+82	+17	-99	31.27	1.01	26.70
Growth	551	738	944	-82	-17	+99	12.20	.39	10.38

X^2 (2)= 81.95 Sig.<.001

Relative Accessibility in 1960 and Change in the Total Number of Manufacturing Establishments 1952 - 1962.

Manufacturing Change	B_1		B_2		B_3		Total	
Decline	168	(35.44)	302	(34.75)	399	(22.69)	869	(28.01)
Growth	306	(64.56)	567	(65.25)	1360	(77.31)	2233	(71.99)
Total	474		869		1759		3102	

	Expected Frequencies			Deviations From Observed Frequencies			Chi-Square Components		
	B_1	B_2	B_3	B_1	B_2	B_3	B_1	B_2	B_3
Decline	133	243	493	+35	+59	-94	9.21	14.32	17.92
Growth	341	626	1266	-35	-59	+94	3.59	5.56	6.98

X^2 (2)= 57.58 Sig.<.001

Change in Relative Accessibility and Change in the Total Number of Manufacturing Establishments 1952 - 1962.

Manufacturing Change	b_1		b_2		Total	
Decline	785	(28.37)	84	(25.08)	869	(28.01)
Growth	1982	(71.63)	251	(74.92)	2233	(71.99)
Total	2767		335		3102	

	Expected Frequencies		Deviations From Observed Frequencies		Chi-Square Components	
	b_1	b_2	b_1	b_2	b_1	b_2
Decline	775	94	+10	-10	.13	1.06
Growth	1992	241	-10	+10	.05	.41

X^2 (1)= 1.65 Sig.<.05

And, turning to the third component of Table 3.3, the greater the growth in national market access, the greater the growth in manufacturing establishments.

Manufacturing growth favor areas of high national market access, and those whose national market access was increasing. However, comparing the two top elements of Table 3.3, the decline in the proportion of high access counties growing between 1950 and 1960, and the increase in manufacturing growth in low-access counties also meant that dispersion was taking place. Not only were the countries able to improve their growth prospects by improving their national market access; industrial growth was apparently slackening somewhat in high access areas -- while still favoring them relatively -- and increasing due to dispersion into peripheral areas.

Decentralization

The third dimension of industrial growth is that of relative centralization/decentralization within FEA's. Table 3.4 shows that growth took place in all local access zones, including areas beyond the reach of daily commuting to FEA centers in 1960 (C_1). However, the proportion of counties at each level of commuting intensity that were growing industrially varied systematically with location, with more of the high-commuting counties growing (86.02%) than the peripheral counties (61.19%). Indeed, more peripheral counties were declining than might be expected based upon the marginal frequencies, and more central counties were growing than expected. Industrial growth favored the high-access locations. Decentralization was clearly taking place, but was highly selective in the more remote peripheries.

Hierarchical Diffusion and Dispersion

So far, we are able to conclude that industrial growth rates increase with hierarchical status of FEA centers, with national mar-

TABLE 3.4

Relative Location in Urban Fields (1960) and Change in the Total Number of Manufacturing Establishments 1952 - 1962.

Manufacturing Change

	C_1	C_2	C_3	C_4	Total P
Decline	255 (38.81)	379 (29.54)	163 (25.19)	72 (13.98)	869 (28.01)
Growth	402 (61.19)	904 (70.46)	484 (74.81)	443 (86.02)	2233 (71.99)
	657	1283	647	515	3102

	Expected Frequencies				Deviations From Observed Frequencies				Chi-Square Components			
	C_1	C_2	C_3	C_4	C_1	C_2	C_3	C_4	C_1	C_2	C_3	C_4
Decline	184	359	181	145	+71	+20	-18	-73	27.40	1.11	1.79	36.75
Growth	473	924	466	370	-71	-20	+18	+73	10.66	.43	.69	14.40

X^2 (2)= 93.23 Sig.$<$ .001

ket access, and with relative location within FEA's. While there were no decadal shifts in growth rates tending to support the idea of hierarchical diffusion of growth, growth was taking place at all levels of the urban hierarchy, although more selectively at lower than higher levels. The growth prospects of counties could be increased by filtering up produced by improvements in the hierarchical status of their FEA centers, however. On the other hand, the growth prospects of more peripheral counties apparently increased in the decade both by improvements in national market access and by dispersion of industrial activity, although growth rates remained lower, and growth prospects therefore more selective, in the peripheral counties at the end of the decade.

The question that now arises is about the nature of this selectivity. Can it be explained by the combinations of relative location shifts? This question will first be explored by looking at the growth of manufacturing establishments in relation to hierarchical status and national market access, considered simultaneously.

The evidence is in Table 3.5, which is similar to Tables 3.2-3.4, except that the dependent variable, growth in industrial establishments, is now placed in the columns, and the various two-way combinations of relative locations are placed in the rows.

Immediately, the finer discriminations are apparent. In fact, growth prospects were greater than expected for high-status FEA's throughout the country in 1950, and for all FEA's in the zone of highest national market access in that year. The growth proportion varied from one county in two for low-status FEA's in peripheral regions to eight counties in ten for counties located in high-status FEA's in the areas of highest national market access. Possessing high potential in 1950 was probably crucial for attracting growth to a high proportion of counties. In the interactions of the two components it seems that the most effective combination was high relative

TABLE 3.5

Urban Hierarchy Interaction With Relative Accessibility 1950 and the Change in the Total Number of Manufacturing Establishments 1952 - 1962

Interaction	Manufacturing Change: Decline		Growth		Total
A_1B_1	230	(46.84)	261	(53.16)	491
A_1B_2	119	(33.90)	233	(66.10)	351
A_1B_3	48	(21.24)	178	(78.76)	266
A_2B_1	51	(30.18)	118	(69.82)	169
A_2B_2	108	(29.35)	260	(70.65)	368
A_2B_3	104	(21.58)	378	(78.42)	482
A_3B_1	16	(15.09)	90	(84.91)	106
A_3B_2	77	(25.16)	229	(74.84)	306
A_3B_3	116	(19.24)	487	(80.76)	603
Total	869	(28.01)	2233	(71.99)	603

	Expected Frequencies: Decline	Growth	Deviations From Observed Frequencies: Decline	Growth	Chi-Square Components: Decline	Growth
A_1B_1	138	353	+92	-92	61.33	23.99
A_1B_2	98	253	+21	-21	4.50	1.74
A_1B_3	63	163	-15	+15	3.57	1.38
A_2B_1	47	122	+ 4	- 4	.34	.13
A_2B_2	103	265	+ 5	- 5	.24	.09
A_2B_3	135	347	-31	+31	7.12	2.77
A_3B_1	30	76	-14	+14	6.53	2.58
A_3B_2	86	220	- 9	+ 9	.94	.37
A_3B_3	169	434	-53	+53	16.62	6.45

X^2 (16)= 140.69 Sig.<.001

Urban Hierarchy Interaction With Relative Accessibility (1960) and the Change in the Total Number of Manufacturing Establishments 1952 - 1962

Interaction	Manufacturing Change: Decline		Growth		Total
A_1B_1	122	(41.9)	169	(58.1)	291
A_1B_2	167	(49.4)	171	(50.6)	338
A_1B_3	75	(23.5)	243	(76.5)	318
A_2B_1	35	(31.2)	77	(68.8)	112
A_2B_2	60	(24.8)	181	(75.2)	241
A_2B_3	145	(22.6)	495	(77.4)	640
A_3B_1	11	(15.4)	60	(84.6)	71
A_3B_2	75	(25.8)	215	(74.2)	290
A_3B_3	179	(22.3)	622	(77.7)	801
Total	869	(28.01)	2233	(71.99)	3102

	Expected Frequencies		Deviations From Observed Frequencies		Chi-Square Components	
A_1B_1	82	209	+40	-40	19.51	7.65
A_1B_2	95	243	+72	-72	54.57	21.33
A_1B_3	89	229	-14	+14	2.20	.85
A_2B_1	31	81	+ 4	- 4	.52	.20
A_2B_2	68	173	- 8	+ 8	.94	.36
A_2B_3	179	461	-34	+34	6.46	2.51
A_3B_1	20	51	+ 9	- 9	4.05	1.59
A_3B_2	81	209	- 6	+ 6	.44	.17
A_3B_3	224	577	-45	+45	9.04	3.44

X^2 (8)= 135.63 Sig.<.001

TABLE 3.5
(cont.)

Interaction Between the Change of F.E.A. City Size and the Change in Relative Accessibility With the Change in the Total Number of Manufacturing Establishments 1952 - 1962.

	Manufacturing Change				
	Decline		Growth		Total
a_1b_1	192	(39.1)	299	(60.9)	491
a_1b_2	51	(44.3)	64	(55.7)	115
a_2b_1	113	(19.8)	456	(80.2)	569
a_2b_2	3	(13.0)	20	(87.0)	23
a_3b_1	209	(25.6)	606	(74.4)	815
a_3b_2	9	(16.3)	46	(83.7)	55
a_4b_1	271	(30.3)	621	(69.7)	892
a_4b_2	21	(14.7)	121	(85.3)	142
Total	869	(28.01)	2233	(71.99)	3102

	Expected Frequencies		Deviations From Expected Frequencies		Chi-Square Components	
a_1b_1	138	353	+54	-54	21.13	8.26
a_1b_2	32	83	+19	-19	11.28	4.31
a_2b_1	159	410	-46	+46	13.31	5.14
a_2b_2	6	17	- 3	+ 3	1.50	.53
a_3b_1	228	587	-19	+19	1.58	.62
a_3b_2	15	40	- 6	+ 6	2.40	.90
a_4b_1	250	642	+21	-21	1.76	.68
a_4b_2	41	101	-20	+20	9.76	3.96

X^2 (7)= 87.12 Sig.$<$.001

accessibility and large urban centers. The first, as has already bee noted, has a positive effect on deviations at all hierarchical levels the latter neutralizes the negative effect of low levels of potentia The extreme deviations are concentrated in the extreme periphery and at the core of the nation's economic space. In fact, almost eighty per cent of the chi-square value is attributed to these two groups (A_1 B_1 and A_3 B_3), and particularly to the first group. This is an indication of the difficulty that the periphery had in attracting growth relative to all other areas in the country during the 1950's.

The evidence for 1960 is the same, although many counties are seen to have improved their growth prospects by both improvements in hierarchical status and national market access. Moreover, growth proportions are seen to be somewhat lower in the high-status high-access locations and to have improved in the low-status peripheral regions.

The reduction in the geographical extent of the periphery is clearly apparent. After the shift of counties from lower relative location groups to upper level groups during the 1950's, the counties that remained in the periphery account for 76 per cent of the sum of the chi-square deviations. These deviations are significantly negative. Although there was growth of manufacturing establishments in a substantial proportion of these counties, they alone contribute most to deviations from the national expectations as to growing and declining counties. All other deviations declined in their contribution to the overall chi-square value. This was probably a result of an effective decline in relative location differentials in so far as growth in total numbers of manufacturing establishments is concerned. One would expect that in the future, industrial change, as measured by this variable, will conform the national distribution of population without any significant deviations because of relative location differentials. What is needed is simply further growth of relative location levels, particularly in the periphery.

The final part of the table, which shows the interaction between change in the central city population and population potential of counties during the 1950's, indicates what could be the cause of negative deviations in all of the tables presented so far, namely the lack of population growth of many peripheral urban centers, regardless of the potential change. It thus seems that in the interaction between the two types of change, FEA central-city population growth is probably the most important in affecting change in the total number of manufacturing establishments. In turn, this implies that the change in the center of an urban field is very important and it possibly points out that what happens elsewhere is dependent on such change.

Hierarchical Diffusion and Decentralization

The pattern that was evident in the interaction between hierarchical diffusion and dispersion repeats itself in the interaction between diffusion and decentralization (Table 3.6). Once again, the operation of simultaneous diffusion and decentralization processes is evident. In no case is the proportion of counties experiencing growth less than about three out of five. However, it is interesting to note that most of negative deviations from expectations are concentrated in outlying counties in the urban fields of small urban centers. On the other hand, the positive deviations, although generally associated with medium-sized and large cities, are most significant closer in to the urban centers of urban fields. Thus, the most pronounced aspect of decentralization is the process of suburbanization at all levels of the urban hierarchy. The higher the level of the urban hierarchy the more accelerated and extensive the process of suburbanization, extending much further out into the largest urban fields. The same pattern repeats itself for the 1960 interactions, where it becomes evident that counties associated with any degree of change in FEA city population, if located in 1960 in the periphery of urban fields, were relatively less likely to exper-ience growth in number of manufacturing establishments. On the oth hand, the opposite is the case when counties are close to the urban center, having a large proportion of their labor force commuting to the central city in 1960. The dominance of accelerated suburbaniza-tion as a sub-process of decentralization is again evident.

Dispersion and Decentralization

Dispersion and decentralization operate simultaneously in what by now is a familiar pattern throughout the spatial-economic system. Table 3.7, describing the proportions of growing and declining coun-ties, ranges once again from one in two counties in areas with low potential in 1950 that were farthest away from the centers of urban fields to nine out of ten in all groups of counties closest to their urban centers, regardless of their relative accessibility in 1950. A clear pattern emerges. Industrial dispersion from areas of high relative accessibility to areas of low relative accessibility is af-fected by relative location within urban fields. For any level of relative accessibility, the closer counties are to their urban cente the greater is the likelihood of experiencing growth. This is parti ularly so for low and medium levels of population potential. High levels of relative accessibility seems to offset the negative effect of location at the periphery of urban fields. In fact, all groups o counties having high population potentials in 1950 show positive de-viations from the national proportion of growing counties.

These conclusions are confirmed for the 1960 pattern of relativ location. The materials on locational change indicate that the closer a county is to its FEA center, the higher the likelihood of growth in manufacturing establishments, regardless of the change in potential. Also, the higher the rate of relative accessibility change,

TABLE 3.6

Interaction Between Urban Hierarchy (1950) and Relative Location in Urban Fields (1960) and the Change in the Total Number of Manufacturing Establishments 1954 - 1964.

Manufacturing Change

	Decline		Growth		Total
A_1 C_1	246	(40.66)	359	(59.34)	605
A_1 C_2	91	(40.26)	135	(59.74)	226
A_1 C_3	42	(32.56)	87	(67.44)	129
A_1 C_4	18	(16.67)	90	(83.33)	108
A_2 C_1	3	(17.65)	14	(82.35)	17
A_2 C_2	154	(30.31)	354	(69.69)	508
A_2 C_3	72	(26.09)	204	(74.91)	276
A_2 C_4	34	(15.60)	184	(84.40)	218
A_3 C_1	6	(17.14)	29	(82.86)	35
A_3 C_2	134	(24.41)	415	(75.59)	549
A_3 C_3	49	(20.25)	193	(79.75)	242
A_3 C_4	20	(10.58)	169	(89.42)	189
Total	869	(28.01)	2233	(71.99)	3102

	Expected Frequencies		Deviations From Observed Frequencies		Chi-Square Components	
A_1 C_1	169	434	+76	-76	34.18	13.31
A_1 C_2	47	122	+21	-21	9.38	3.61
A_1 C_3	27	71	+ 8	- 8	2.37	.90
A_1 C_4	22	55	- 6	+ 6	1.64	.65
A_2 C_1	5	13	- 2	+ 2	.80	.31
A_2 C_2	137	351	+ 8	- 8	.47	.18
A_2 C_3	75	194	-11	+11	1.61	.62
A_2 C_4	61	157	-33	+33	17.85	6.94
A_3 C_1	10	26	- 3	+ 3	.90	.35
A_3 C_2	175	451	- 9	+ 9	.46	.18
A_3 C_3	78	202	-14	+14	2.51	.97
A_3 C_4	63	157	-35	+35	19.44	7.80

X^2 (22) = 127.43 Sig. < .001

TABLE 3.6
(cont.)

Interaction Between Urban Hierarchy (1960) and Relative Location in Urban Fields (1960) With the Change in the Total Number of Manufacturing Establishments 1952 - 1962.

	Manufacturing Change				
	Decline		Growth		Total
A_1 C_1	245	(40.63)	358	(59.37)	603
A_1 C_2	68	(40.24)	101	(59.76)	169
A_1 C_3	35	(35.71)	63	(64.29)	98
A_1 C_4	16	(20.78)	61	(79.22)	77
A_2 C_1	3	(16.67)	15	(83.33)	18
A_2 C_2	145	(29.71)	343	(80.29)	488
A_2 C_3	64	(23.79)	205	(76.21)	269
A_2 C_4	28	(12.84)	190	(87.16)	218
A_3 C_1	7	(19.44)	29	(80.56)	36
A_3 C_2	166	(26.52)	460	(73.48)	626
A_3 C_3	64	(22.86)	216	(77.14)	280
A_3 C_4	28	(12.73)	192	(87.27)	220
Total	869	(28.01)	2233	(71.99)	3102

	Expected Frequencies		Deviations From Observed Frequencies		Chi-Square Components	
A_1 C_1	169	436	+77	-77	35.08	13.60
A_1 C_2	63	163	+28	-28	12.44	4.81
A_1 C_3	36	93	+ 6	- 6	1.00	.39
A_1 C_4	30	88	-12	+12	4.80	1.64
A_2 C_1	5	12	- 2	+ 2	.80	.33
A_2 C_2	142	366	+12	-12	1.01	.39
A_2 C_3	77	199	- 5	+ 5	.32	.13
A_2 C_4	61	157	-27	+27	11.95	4.64
A_3 C_1	10	25	- 4	+ 4	1.60	.64
A_3 C_2	154	395	-20	+20	2.60	1.01
A_3 C_3	68	174	-19	+19	5.31	2.07
A_3 C_4	54	135	-34	+34	21.41	8.56

χ^2 (20)= 136.73 Sig <.001

TABLE 3.6
(cont.)

Interaction Between Population Change of Centers in the Urban Hierarchy 1950 - 1960 and Relative Location in Urban Fields (1960) with Change in the Total Number of Manufacturing Establishments.

Interaction	Manufacturing Change: Decline		Manufacturing Change: Growth		Total
$a_1 + a_2\ C_{1+2}$	312	(33.95)	607	(66.05)	919
$a_1 + a_2\ C_3$	27	(18.12)	122	(81.88)	149
$a_1 + a_2\ C_4$	20	(15.38)	110	(74.62)	130
$a_3\ C_{1+2}$	121	(28.34)	306	(71.66)	427
$a_3\ C_3$	71	(28.51)	178	(71.49)	249
$a_3\ C_4$	26	(13.40)	168	(86.60)	194
$a_4\ C_{1+2}$	201	(33.84)	393	(66.16)	594
$a_4\ C_3$	65	(26.10)	184	(73.90)	249
$a_4\ C_4$	26	(13.61)	165	(86.39)	191
Total	869	(28.01)	2233	(71.99)	3102

	Expected Frequencies		Deviations From Observed Frequencies		Chi-Square Components	
$a_1 + a_2\ C_{1+2}$	257	662	+55	-55	11.77	4.57
$a_1 + a_2\ C_3$	42	107	-15	+15	5.36	2.10
$a_1 + a_2\ C_4$	36	94	-16	+16	7.11	2.72
$a_3\ C_{1+2}$	120	307	+ 1	- 1	--	--
$a_3\ C_3$	70	179	+ 1	- 1	.01	--
$a_3\ C_4$	54	140	-28	+28	14.52	5.60
$a_4\ C_{1+2}$	166	428	+35	-35	22.68	2.86
$a_4\ C_3$	70	179	- 5	+ 5	.36	.14
$a_4\ C_4$	54	137	-28	+28	14.52	5.72

$X^2\ (16) = 100.04$ Sig. $< .001$

TABLE 3.7

Interaction Between Relative Accessibility (1950) and Relative Location Within Urban Field (1960) With Change in the Total Number of Manufacturing Establishments 1952 - 1962.

	Manufacturing Change				
	Decline		Growth		Total
B_1C_1	165	(49.25)	170	(50.75)	335
B_1C_2	94	(37.60)	156	(62.40)	250
B_1C_3	27	(32.14)	57	(67.86)	84
B_1C_4	11	(11.34)	86	(88.66)	97
B_2C_1	61	(34.08)	118	(65.92)	179
B_2C_2	153	(31.81)	328	(68.19)	481
B_2C_3	69	(31.08)	153	(68.92)	222
B_2C_4	21	(14.68)	122	(85.32)	143
B_3C_1	29	(20.28)	114	(79.72)	143
B_3C_2	132	(23.91)	420	(76.19)	552
B_3C_3	67	(19.65)	274	(80.35)	341
B_3C_4	40	(14.54)	235	(85.46)	275
Total	869	(28.01)	2233	(71.99)	3102

	Expected Frequencies		Deviations From Expected Frequencies		Chi-Square Components	
B_1C_1	94	341	+71	-71	53.63	14.78
B_1C_2	70	180	+24	-24	8.23	3.20
B_1C_3	24	60	+ 3	- 3	.38	.15
B_1C_4	27	70	-16	+16	9.48	3.66
B_2C_1	50	129	+11	-11	2.42	.94
B_2C_2	135	346	+18	-18	2.40	.94
B_2C_3	62	160	+ 7	- 7	.79	.31
B_2C_4	40	103	-19	+19	9.02	3.50
B_3C_1	40	103	-11	+11	3.02	1.17
B_3C_2	155	397	-23	+23	3.41	1.33
B_3C_3	96	245	-29	+29	8.76	3.43
B_3C_4	76	199	-36	+36	17.05	6.51

X^2 (22)= 158.51 Sig.< .001

TABLE 3.7
(cont.)

Interaction Between Relative Accessibility (1960) and Relative Location Within Urban Fields (1960) With Change in the Total Number of Manufacturing Establishments (1952 - 1962).

	Manufacturing Change				
	Decline		Growth		Total
$B_1 C_1$	101	(43.16)	133	(56.84)	234
$B_1 C_2$	47	(33.33)	94	(66.67)	141
$B_1 C_3$	14	(29.79)	33	(70.21)	47
$B_1 C_4$	6	(11.54)	46	(88.46)	52
$B_2 C_1$	105	(53.03)	93	(46.97)	198
$B_2 C_2$	135	(33.50)	268	(66.50)	403
$B_2 C_3$	47	(31.76)	101	(68.24)	148
$B_2 C_4$	15	(12.50)	105	(87.50)	120
$B_3 C_1$	49	(21.78)	176	(78.22)	225
$B_3 C_2$	197	(26.66)	542	(73.34)	739
$B_3 C_3$	102	(22.57)	350	(77.43)	452
$B_3 C_4$	51	(14.87)	292	(85.13)	343
	869	(28.01)	2233	(71.99)	3102

	Expected Frequencies		Deviations From Expected Frequencies		Chi-Square Components	
$B_1 C_1$	66	168	+35	-35	18.56	7.29
$B_1 C_2$	39	102	+ 8	- 8	1.64	.63
$B_1 C_3$	13	34	+ 1	- 1	.08	.03
$B_1 C_4$	15	37	- 9	+ 9	5.40	2.19
$B_2 C_1$	55	143	+50	-50	45.45	17.48
$B_2 C_2$	113	290	+22	-22	4.28	1.67
$B_2 C_3$	41	107	+ 6	-	.88	.34
$B_2 C_4$	34	86	-19	+19	10.62	4.20
$B_3 C_1$	63	162	-14	+14	3.11	1.21
$B_3 C_2$	207	532	-10	+10	.48	.19
$B_3 C_3$	127	325	-25	+25	4.92	1.92
$B_3 C_4$	96	247	-45	+45	21.09	8.20

X^2 (22)= 161.86 Sig.<.001

TABLE 3.7
(cont.)

Interaction Between Change in Relative Accessibility 1950 - 1960 and Relative Location Within Urban Fields (1960) With Change in the Total Number of Manufacturing Establishments 1952 - 1962.

Interaction	Manufacturing Change				
	Decline		Growth		Total
b_1 C_1	203	(37.94)	332	(62.06)	535
b_1 C_2	359	(29.97)	839	(70.03)	1198
b_1 C_3	153	(25.37)	450	(74.63)	603
b_1 C_4	70	(16.24)	361	(83.76)	431
b_2 C_1	52	(42.62)	70	(57.38)	122
b_2 C_2	20	(23.53)	65	(76.47)	85
b_2 C_3	10	(22.73)	34	(77.23)	44
b_2 C_4	2	(2.44)	82	(97.56)	84
Total	869	(28.01)	2233	(71.99)	3102

	Expected Frequencies		Deviations From Expected Frequencies		Chi-Square Components	
b_1C_1	150	385	+53	-53	18.73	7.30
b_1C_2	336	862	+23	-23	1.57	.61
b_1C_3	169	434	-16	+16	1.51	.59
b_1C_4	121	310	-51	+51	21.49	8.39
b_2C_1	34	88	+18	-18	9.53	3.68
b_3C_2	24	61	- 4	+ 4	.67	.26
b_2C_3	12	32	- 2	+ 2	.33	.12
b_2C_4	23	61	-21	+21	19.17	7.23

X^2 (13)= 101.18 Sig.<.001

the greater the likelihood of positive manufacturing change, regardless of location in the urban field.

Relative Location and Rate of Change Differentials

So far, we have considered the relationships between growth or decline in numbers of manufacturing establishments and the relative location of counties within the nation's space-economy. Now we turn to the actual rates of change of manufacturing establishments, using analysis of variance to determine which components of relative location are most significantly related to the manufacturing rate of change differentials, and least-squares covariance estimates to predict the rate of manufacturing growth. Table 3.8 shows the group and subgroup manufacturing growth rates that were analyzed.

The first analysis of variance that was performed used a three-way model in which the classifications of relative location were entered in the conventional way, first the three "main" effects A, B and C; second, the first-order interactions AB, AC and BC; and finally the three-way interaction ABC. As the uppermost part of Table 3.9 indicates, when arranged in that order, only three components of relative location are statistically significant: relative accessibility in 1950 - B, interaction of hierarchical status and relative accessibility - AB, and interaction of national and local market access - BC.

Only these three statistically-significant effects were entered into the third phase of the analysis, an analysis with the purpose of determining the significance, if any, of each effect when others are held constant statistically. The results after introducing the effects into the analysis in the order of their statistical significance in the second part of Table 3.9 shows that two effects are the dominant ones, the interaction of relative accessibility with relative

TABLE 3.8

Observed Means of Major Dimensions and Their Interaction: Change in Total Number of Manufacturing Establishments 1952 - 1962.

F.E.A. City Population 1950 (A)		N	$\bar{X}$
Small	(A_1)	1068	1.42
Medium	(A_2)	1019	1.62
Large	(A_3)	1015	1.57

Population Potential 1960 (B)		N	$\bar{X}$
Low	(B_1)	766	1.75
Medium	(B_2)	1025	1.40
High	(B_3)	1311	1.52

Commuting Levels 1960 (C)		N	$\bar{X}$
No Commuting	(C_1)	657	1.35
Low	(C_2)	1283	1.59
Medium	(C_3)	647	1.44
High	(C_4)	515	1.77

Interaction AB	N	$\bar{X}$
A_1B_1	491	1.35
A_1B_2	351	1.40
A_1B_3	226	1.64
A_2B_1	169	2.30
A_2B_2	368	1.34
A_2B_3	482	1.60
A_3B_1	106	2.75
A_3B_2	306	1.48
A_3B_3	603	1.42

Interaction AC	N	$\bar{X}$
A_1C_1	605	1.34
A_1C_2	226	1.54
A_1C_3	129	1.45
A_1C_4	108	1.59
A_2C_1	17	1.54
A_2C_2	508	1.67
A_2C_3	276	1.46
A_2C_4	218	1.74
A_3C_1	35	1.34
A_3C_2	549	1.54
A_3C_3	242	1.42
A_3C_4	189	1.90

Interaction BC	N	$\bar{X}$
B_1C_1	335	1.27
B_1C_2	250	1.83
B_1C_3	84	1.45
B_1C_4	97	3.45
B_2C_1	179	1.40
B_2C_2	481	1.38
B_2C_3	222	1.40
B_2C_4	143	1.46
B_3C_1	143	1.48
B_3C_2	552	1.66
B_3C_3	341	1.47
B_3C_4	275	1.34

Interaction ABC

	N	$\bar{X}$	S
$A_1B_1C_1$	329	1.26	1.32
$A_1B_1C_2$	80	1.34	1.53
$A_1B_1C_3$	39	1.53	1.02
$A_1B_1C_4$	43	1.86	1.16
$A_1B_2C_1$	168	1.40	.88
$A_1B_2C_2$	94	1.37	.81
$A_1B_2C_3$	50	1.42	.91
$A_1B_2C_4$	39	1.41	.53
$A_1B_3C_1$	108	1.53	.79
$A_1B_3C_2$	52	2.16	6.26
$A_1B_3C_3$	40	1.41	.47
$A_1B_3C_4$	26	1.41	.45

	N	$\bar{X}$	S
$A_2B_1C_1$	3	2.33	1.26
$A_2B_1C_2$	105	1.93	3.87
$A_2B_1C_3$	24	1.27	1.01
$A_2B_1C_4$	37	3.99	13.92
$A_2B_2C_1$	5	1.41	.54
$A_2B_2C_2$	206	1.37	1.59
$A_2B_2C_3$	93	1.29	.45
$A_2B_2C_4$	64	1.29	.40
$A_2B_3C_1$	9	1.34	.44
$A_2B_3C_2$	197	1.83	5.65
$A_2B_3C_3$	159	1.58	2.98
$A_2B_3C_4$	117	1.28	1.17

	N	$\bar{X}$	S
$A_3B_1C_1$	3	1.50	.18
$A_3B_1C_2$	65	2.27	4.07
$A_3B_1C_3$	21	1.52	.60
$A_3B_1C_4$	17	6.30	19.15
$A_3B_2C_1$	6	1.41	.19
$A_3B_2C_2$	181	1.40	.62
$A_3B_2C_3$	79	1.52	.82
$A_3B_2C_4$	40	1.77	.79
$A_3B_3C_1$	26	1.31	.39
$A_3B_3C_1$	303	1.47	2.82
$A_3B_3C_3$	142	1.36	.45
$A_3B_3C_4$	132	1.37	.51

TABLE 3.8

cont.

F.E.A City Population 1960 (A)		N	$\overline{X}$
Small	(A_1)	947	1.40
Medium	(A_2)	993	1.66
Large	(A_3)	1162	1.55

Population Potential 1960 (B)		N	$\overline{X}$
Low M	(B_1)	474	1.58
Medium	(B_2)	869	1.65
High	(B_3)	1759	1.48

Commuting Levels 1960 (C)		N	$\overline{X}$
No Commuting	(C_1)	657	1.35
Low	(C_2)	1283	1.59
Medium	(C_3)	647	1.44
High	(C_4)	515	1.77

Interaction AB	N	$\overline{X}$
A_1B_1	291	1.43
A_1B_2	338	1.20
A_1B_3	318	1.59
A_2B_1	112	1.92
A_2B_2	241	1.96
A_2B_3	640	1.50
A_3B_1	71	1.65
A_3B_2	290	1.90
A_3B_3	801	1.41

Interaction AC	N	$\overline{X}$
A_1C_1	603	1.35
A_1C_2	169	1.58
A_1C_3	98	1.38
A_1C_4	77	1.50
A_2C_1	18	1.54
A_2C_2	488	1.67
A_2C_3	269	1.52
A_2C_4	218	1.82
A_3C_1	36	1.32
A_3C_2	626	1.53
A_3C_3	280	1.40
A_3C_4	220	1.82

Interaction BC	N	$\overline{X}$
B_1C_1	234	1.40
B_1C_2	141	1.93
B_1C_3	47	1.40
B_1C_4	52	1.60
B_2C_1	198	1.17
B_2C_2	403	1.54
B_2C_3	148	1.48
B_2C_4	120	3.00
B_3C_1	225	1.46
B_3C_2	739	1.56
B_3C_3	452	1.44
B_3C_4	343	1.36

Interaction ABC

	N	$\overline{X}$	S		N	$\overline{X}$	S		N	$\overline{X}$	S
$A_1B_1C_1$	230	1.40	1.49	$A_2B_1C_1$	1	1.00	--	$A_3B_1C_1$	3	1.50	.18
$A_1B_1C_2$	26	1.50	1.35	$A_2B_1C_2$	69	2.23	4.79	$A_3B_1C_2$	46	1.71	1.16
$A_1B_1C_3$	18	1.54	.78	$A_2B_1C_3$	19	1.23	.55	$A_3B_1C_3$	10	1.48	.61
$A_1B_1C_4$	17	1.64	.78	$A_2B_1C_4$	23	1.57	.59	$A_3B_1C_4$	12	1.61	.63
$A_1B_2C_1$	191	2.14	.88	$A_2B_2C_1$	4	2.42	.80	$A_3B_2C_1$	3	1.36	.20
$A_1B_2C_2$	83	1.22	.86	$A_2B_2C_2$	134	1.54	2.01	$A_3B_2C_2$	186	1.67	2.49
$A_1B_2C_3$	32	1.13	.74	$A_2B_2C_3$	48	1.61	1.00	$A_3B_2C_3$	68	1.56	.89
$A_1B_2C_4$	32	1.60	.85	$A_2B_2C_4$	55	3.22	11.45	$A_3B_2C_4$	33	3.99	13.76
$A_1B_3C_1$	182	1.50	.74	$A_2B_3C_1$	13	1.31	.40	$A_3B_3C_1$	30	1.30	.39
$A_1B_3C_2$	60	2.09	5.82	$A_2B_3C_2$	285	1.59	4.64	$A_3B_3C_2$	394	1.45	2.58
$A_1B_3C_3$	48	1.48	.81	$A_2B_3C_3$	202	1.52	2.65	$A_3B_3C_3$	202	1.34	.46
$A_1B_3C_4$	28	1.28	.41	$A_2B_3C_4$	140	1.31	1.07	$A_3B_3C_4$	175	1.42	.59

TABLE 3.8

(cont.)

F.E.A. City Population Change (a)

	N	$\bar{X}$
No Change	606	1.35
Decline	592	1.49
Growth	870	1.63
Rapid Growth	1034	1.60

County Population Potential Change (a)

	N	$\bar{X}$
Growth	2767	1.46
Rapid Growth	335	2.12

Interaction ab

	N	$\bar{X}$
a_1 b_1	491	1.34
a_1 b_2	115	1.37
a_2 b_1	569	1.48
a_2 b_2	23	1.94
a_3 b_1	815	1.52
a_3 b_2	55	3.19
a_4 b_1	892	1.47
a_4 b_2	142	2.41

Interaction aC

	N	$\bar{X}$
a_1 C_1	601	1.35
a_1 C_2	3	1.24
a_1 C_3	1	1.08
a_1 C_4	1	1.22
a_2 C_1	21	1.39
a_2 C_2	294	1.57
a_2 C_3	148	1.48
a_2 C_4	129	1.34
a_3 C_1	16	1.24
a_3 C_2	411	1.64
a_3 C_3	249	1.45
a_3 C_4	194	1.86
a_4 C_1	19	1.46
a_4 C_2	575	1.56
a_4 C_3	249	1.41
a_4 C_4	191	1.97

Interaction bC

	N	$\bar{X}$
b_1 C_1	535	1.34
b_1 C_2	1198	1.58
b_1 C_3	603	1.43
b_1 C_4	431	1.34
b_2 C_1	122	1.40
b_2 C_2	85	1.69
b_2 C_3	44	1.61
b_2 C_4	84	3.97

Interaction abC	N	$\bar{X}$	S
$a_1b_1C_1$	486	1.34	1.18
$a_1b_1C_2$	3	1.24	.12
$a_1b_1C_3$	1	1.08	-
$a_1b_1C_4$	1	1.22	-
$a_1b_2C_1$	115	1.37	.95
$a_1b_2C_2$	-	-	-
$a_1b_2C_3$	-	-	-
$a_1b_2C_4$	-	-	-
$a_2b_1C_1$	21	1.39	.40
$a_2b_1C_2$	289	1.58	2.91
$a_2b_1C_3$	141	1.46	.86
$a_2b_1C_4$	118	1.25	.39
$a_2b_2C_1$	-	-	-
$a_2b_2C_2$	5	1.21	.80
$a_2b_2C_3$	7	1.96	.74
$a_2b_2C_4$	11	2.27	.54
$a_3b_1C_1$	15	1.21	.43
$a_3b_1C_2$	394	1.63	4.56
$a_3b_1C_3$	230	1.45	2.46
$a_3b_1C_4$	176	1.40	1.05
$a_3b_2C_1$	1	1.71	-
$a_3b_2C_2$	17	1.86	1.55
$a_3b_2C_3$	19	1.46	.57
$a_3b_2C_4$	18	6.36	18.54
$a_4b_1C_1$	13	1.20	.30
$a_4b_1C_2$	512	1.55	2.62
$a_4b_1C_3$	231	1.39	.68
$a_4b_1C_4$	136	1.34	.48
$a_4b_2C_1$	6	2.01	.90
$a_4b_2C_2$	63	1.68	1.05
$a_4b_2C_3$	18	1.64	1.05
$a_4b_2C_4$	55	3.53	11.41

TABLE 3.9

Analysis of Variance (1950)

Effect	Mean Square	Degrees of Freedom	F. Ratio	Significance
A	11.30	2, 3066	1.17	<.3101
B	41.43	2, 3066	4.28	<.0139
C	15.27	3, 3066	1.58	<.1910
AB	47.43	4, 3066	4.90	<.0007
AC	1.78	6, 3066	.18	<.9816
BC	52.36	6, 3066	5.41	<.0001
ABC	9.70	12, 3066	1.00	<.4414

Dependent Variable Variance= 9.68 With 3066 Degrees of Freedom

ANOVA: Each Effect Alone

Effect	Mean Square	Degrees of Freedom	F. Ratio	Significance
A	11.30	2, 3066	1.17	<.3101
B	27.07	2, 3066	2.80	<.0608
C	20.03	3, 3066	2.07	<.1011
AB	24.77	4, 3066	2.56	<.0366
AC	5.00	6, 3066	.52	<.7967
BC	31.40	6, 3066	3,24	<.0035
ABC	5.06	12, 3066	.52	<.9020

ANOVA: Significant Effects Only

	Effect	Mean Square	Degrees of Freedom	F. Ratio	Significance
1.	BC	31.40	6, 3066	3.24	<.0035
	B	57.89	2, 3066	5.98	<.0026
2.	B	27.07	2, 3066	2.80	<.0608
	BC	41.68	6, 3066	4.31	<.0003

Least Square Estimates of the Location Types Differing Significantly from the General Mean

General Mean	+1.61
B_1	+ .42
B_2	- .14
B_3	+ .28
B_1C_1	-2.05
B_1C_2	- .03
B_1C_3	-1.51
B_1C_4	+3.59
B_2C_1	- .17
B_2C_2	-1.44
B_2C_3	+ .08
B_2C_4	+1.32

TABLE 3.9

Estimated Group Means and Residuals in the Form of T-statistics

	Means	Residuals
A_1 B_1 C_1	1.40	-0.04
A_1 B_1 C_2	1.81	-0.15
A_1 B_1 C_3	1.77	-0.07
A_1 B_1 C_4	2.76	-0.28
A_1 B_2 C_1	1.64	-0.08
A_1 B_2 C_2	1.37	-0.00
A_1 B_2 C_3	1.52	-0.03
A_1 B_2 C_4	0.98	0.13
A_1 B_3 C_1	1.78	-0.08
A_1 B_3 C_2	1.65	0.16
A_1 B_3 C_3	1.54	-0.04
A_1 B_3 C_4	1.09	0.10
A_2 B_1 C_1	1.40	0.29
A_2 B_1 C_2	1.81	0.03
A_2 B_1 C_3	1.77	-0.16
A_2 B_1 C_4	2.76	0.39
A_2 B_2 C_1	1.64	-0.07
A_2 B_2 C_2	1.37	0.00
A_2 B_2 C_3	1.52	-0.07
A_2 B_2 C_4	0.98	0.09
A_2 B_3 C_1	1.78	-0.14
A_2 B_3 C_2	1.65	0.05
A_2 B_3 C_3	1.54	0.00
A_2 B_3 C_4	1.09	0.06
A_3 B_1 C_1	1.40	0.03
A_3 B_1 C_2	1.81	0.14
A_3 B_1 C_3	1.77	-0.08
A_3 B_1 C_4	2.76	1.13
A_3 B_2 C_1	1.64	-0.07
A_3 B_2 C_2	1.37	0.00
A_3 B_2 C_3	1.52	-0.00
A_3 B_2 C_4	0.98	0.25
A_3 B_3 C_1	1.78	-0.15
A_3 B_3 C_2	1.65	-0.05
A_3 B_3 C_3	1.54	-0.05
A_3 B_3 C_4	1.09	0.08

D. F. = 3066

Residuals Estimated After Fitting as Model of Rank 9

location within the urban field (BC), and relative accessibility as such (B). The results imply that given relative location in 1950, the most important elements of location in the national space-economy affecting change in total number of manufacturing establishments were relative national market access together with its interaction with the location of counties within urban fields. The fourth part of Table 3.9 gives least-squares estimates of the growth-rate differences for the statistically-significant locational subgroups, while the final part presents the full-array of three-way subgroup growth rates predicted from relative national and local market access alone.

Other analyses should have then followed by incorporating the types of relative location that are statistically insignificant into the within-group variation. Instead, to save what appeared to be an unnecessary effort, a second analysis of variance was done separately six times, hypothesizing that each main or principal two-way interaction effect was the most important. The results tended to verify what was discovered in the first analysis, namely that only the three effects, B, AB, and BC contribute to the explanation of change of industrial plants in a statistically significant way, as the second sub-table in Table 3.9 indicates. To emphasize, in this sub-table each effect is entered alone into the analysis, ignoring all others. It implies (in anology with regression analysis) that the three effects: B, AB, and BC have statistically significant simple correlations with the change in the total number of manufacturing establishments.

Tables 3.10-3.12 show similar analysis of variance and covariance materials for manufacturing growth rates related to county locations in 1960, locational changes 1950-1960, and the combined effects of 1950 location and 1950-1960 locational change. In 1960 no main effect was significant; instead, the most significant effects were the interaction of national and local market access (BC) and of

TABLE 3.10

Analysis of Variance, 1960.

Unpooled Analysis of Variance

Effect	Mean Square	Degrees of Freedom	F. Ratio	Significance
A	15.89	2 ,3066	1.62	<.1958
B	16.70	2 3066	1.71	<.1719
C	12.46	3 ,3066	1.28	<.2782
AB	22.79	4 ,3066	2.33	<.0534
AC	1.67	6 ,3066	.17	<.9848
BC	41.04	6 ,3066	4.20	<.0004
ABC	3.53	12 ,3066	.36	<.9767

The dependent variable variance is 9.77

ANOVA: Each Effect Alone

Effect	Mean Square	Degrees of Freedom	F. Ratio	Significance
A	15.89	2 ,3006	1.63	<.1958
B	8.83	2 ,3066	.90	<.4091
C	20.03	3 ,3066	2.05	<.1038
AB	21.46	4 ,3066	2.20	<.0665
AC	5.05	6 ,3066	.52	<.7967
BC	30.23	6 ,3066	3.09	<.0051
ABC	4.79	12 ,3066	.49	<.9219

ANOVA: Significant Effects Only

	Effect	Mean Square	Degrees of Freedom	F. Ratio	Significance
1.	BC	30.23	6 ,3066	3.09	<.0051
	AB	15.26	4 ,3066	1.56	<.1805
2.	AB	21.46	4 ,3006	2.20	<.0665
	BC	26.09	6 ,3066	2.67	<.0138

Least Square Estimates of Effects

General Mean	+1.57
B_1C_1	+ .15
B_1C_2	-1.62
B_1C_3	+ .56
B_2C_1	+1.20
B_2C_2	+ .66
B_2C_3	-1.24

TABLE 3.11

Analysis of Variance (1950-1960)

Unpooled Analysis of Variance

Effect	Mean Square	Degrees of Freedom	F. Ratio	Significance
a	11.48	3, 3074	1.19	<.3110
b	157.52	1, 3074	16.28	<.0001
C	7.61	3, 3074	.79	<.5041
ab	32.00	3, 3074	3.31	<.0193
aC	2.41	9, 3074	.25	<.9871
bC	96.05	3, 3074	9.92	<.0001
abC	N.A.		N.A.	

The dependent variable is 9.68

ANOVA: Each Effect Alone

Effect	Mean Square	Degrees of Freedom	F. Ratio	Significance
a	11.48	3, 3074	1.19	<.3110
b	139.54	1, 3074	14.42	<.0002
C	20.03	3, 3074	2.07	<.1011
ab	4.10	3, 3074	.42	<.7373
aC	9.43	9, 3074	.97	<.4608
bC	28.70	3, 3074	2.96	<.0306
abC	18.10	9, 3074	1.35	<.2021

ANOVA: Significant Effects Only

Effect	Mean Square	Degrees of Freedom	F. Ratio	Significance
1. b	139.54	1, 3074	14.42	<.0002
bC	36.60	3, 3074	3.78	<.0101
2. bC	28.70	3, 3074	2.96	<.0306
b	163.24	1, 3074	16.87	<.0001

Least Square Estimates of Effects

General Mean	+1.78
b_1	- .75
b_1C_1	+ .87
b_1C_2	+1.09
b_1C_3	+ .82

TABLE 3.12

Analysis of Variance: 1950 Relative Location and 1950 - 1960 Locational Change

Unpooled Analysis of Variance

Effect	Mean Square	Degrees of Freedom	F. Ratio	Significance
Aa	7.41	6, 2959	.77	.5956
Bb	7.25	2, 2959	.75	.4730
C	10.78	3, 2959	1.11	.3397
AaBb	7.30	12, 2959	.76	.6970
AaC	7.96	18, 2959	.82	.6756
BbC	20.29	6, 2959	2.10	.0502

The dependent variable variance is 9.67.

ANOVA: Each Effect Alone

Effect	Mean Square	Degrees of Freedom	F. Ratio	Significance
Aa	7.41	6, 2959	.77	.5956
Bb	8.41	2, 2959	.87	.4205
C	20.03	3, 2959	2.07	.1014
AaBb	7.19	12, 2959	.74	.7088
AaC	9.59	18, 2959	.99	.4665
BbC	19.68	6, 2959	2.04	.0576

Least Squares Estimates of Effects

General Mean	1.56
$B_1b_1C_1$	1.14
$B_1b_1C_2$	1.54
$B_1b_1C_3$	2.07
$B_2b_1C_1$	.91
$B_2b_1C_2$	1.34
$B_2b_1C_3$	.90

hierarchical status and national market access (AB). Between 1950 and 1960, the most significant relationships of manufacturing growth rates were to changes in national market access (b), and their interactions with local market access (bC). Finally, considering both initial relative location and changes in relative location, the most significant effect is the three-way combined interaction of changes in hierarchical status, changes in national market access, and local market access.

SUMMARY OF FINDINGS

Drawing together all of these materials, a variety of conclusions emerge about the spatial components of changes in the number of manufacturing establishments during the 1950's:

1. In one sense, diffusion, dispersion and decentralization took place in that growth took place at all levels of the hierarchy, and in all zones of national and local market access. However, the probability that a county would grow increased with hierarchical status and with both national and local market access, or, to express the converse: the lower the status of its FEA and the more peripheral its location, the lower the probability that a county would be numbered among those growing the decade.

2. Over the period of the decade, the pattern of hierarchical diffusion did not change; growth probabilities remained unchanged for each echelon of the urban hierarchy. However, many counties improved their growth prospects in the decade by improving their hierarchical status. In this, counties "filtered up" rather than new growth prospects "filtering down."

3. On the other hand, while the spatial extent of the nation's heartland increased through improvements in national market access, thereby improving the growth prospects of the new heartland counties, there is clear evidence of increased dispersion because during the decade the growth prospects of heartland counties was reduced somewhat while that of hinterland counties improved.

4. These effects interacted. Within the zone of lowest national market access, growth remained concentrated in the highest-status FEA's, and within them in the innermost zones of greatest local market access. At the other extreme, within the heartland, growth took place at all levels of the urban hierarchy, and at successively higher levels decentralized ever further out within the urban field. Dispersion was most evident in the improved growth prospects of medium-status FEA's in the more accessible of the nation's hinterland regions.

5. The most consistently significant and general source of change and of growth rate differentials in the decade was national market access, thus confirming Harris's speculations. Hierarchical status and local market access tended to be factors whose significance varied from one zone of national market access to another, and therefore assumed significance only in their interactions with the more general variable.

These conclusions only apply to one dependent variable -- change in the total number of manufacturing establishments. The purpose of the next three chapters will therefore be to examine three pairs of dependent variables in a similar manner: changes in place of residence and place of work in Chapter 4; changes in numbers of medium- and large-scale manufacturing establishments in Chapter 5; and change in payrolls and value-added in Chapter 6.

CHAPTER 4

CHANGES IN PLACE OF RESIDENCE AND PLACE OF WORK

This chapter and the two to follow are composed of six identically-structured sections. Each dependent variable is analyzed using the methodology outlined in Chapter 3, but since the sequence is in some sense cumulative, many of the things that are said will inevitably be repetitive. The attempt to draw together the full set of conclusions into an organized whole is reserved for Chapter 7.

CHANGES IN PLACE OF RESIDENCE

There are many possible reasons for changes in number of manufacturing employees by place of residence. First is changing mobility. It is very likely that much of the change in place of residence is not an indication of industrialization, but of industrial employees moving within metropolitan areas and from one city to another. Second, the place of residence variable also includes a significant portion of service employees associated with manufacturing research and development and administrative offices. The location of these and other auxiliary services often differs from the location of manufacturing plants. Finally, the variable can represent genuine changes in manufacturing employment arising from industrialization of rural counties. This complex mixture dictates caution in interpreting the association between relative location and change in manufacturing employment by place of residence.

Hierarchical Diffusion

Hierarchical diffusion is evident and strong (Table 4.1). The highest proportion of growing counties is not in high-status FEA's, but in those of lowest status: seven out of ten counties associated with small FEA cities grew in both 1950 and in 1960. This is a clear indication that the periphery is catching up. Since there was not any similar increase in number of establishments (nor, as we shall see, in employment by place of work) it implies that such growth must represent either an increase in commuting to counties in higher-status urban fields from those in urban fields of lower status, or alternatively, increasingly extensive commuting to growing industrial establishments in low-status cities. However, as the third part of Table 4.1 shows, the change in FEA city population is not associated with commuting change. The growth proportion is uniformly distributed in all categories of change in FEA city population.

Dispersion

A trend to dispersion of industrial workers becomes even more evident from an examination of Table 4.2. This table relates the population potential of counties to change in manufacturing employment by place of residence. The chi-square results show a significant statistical association between the two variables. Absolute dispersion is manifested by the fact that both in 1950 and in 1960, more than 60 per cent of all counties experienced growth in manufacturing employment by place of residence. In terms of relative dispersion the medium-level access counties experienced the most growth. Note the high positive deviation of counties with the lowest level of relative accessibility, and the negative deviation of the counties with high relative accessibility, supporting the interpretation of an active dispersion of manufacturing employment by place of residence during the 1952-1962 period.

TABLE 4.1

Urban Hierarchy (1950) and Change in Number of Manufacturing Employees by Place of Residence 1950 - 1960.

Manufacturing Change	A_1		A_2		A_3		Total	
Decline	319	(29.87)	418	(41.02)	422	(41.58)	1159	(37.36)
Growth	749	(70.13)	601	(58.98)	593	(58.42)	1943	(62.64)
Total	1068		1019		1015		3102	

	Expected Frequencies			Deviations From Observed Frequencies			Chi-Square Components		
	A_1	A_2	A_3	A_1	A_2	A_3	A_1	A_2	A_3
Decline	399	381	379	-80	+37	+43	16.04	3.59	4.88
Growth	669	638	636	+80	-37	-43	9.57	2.15	2.91

X^2 (2)= 39.14 Sig.<.001

Urban Hierarchy (1960) and Change in Number of Manufacturing Employees by Place of Residence 1950 - 1960.

Manufacturing Change	A_1		A_2		A_3		Total	P
Decline	266	(28.09)	404	(40.69)	489	(42.09)	1159	(37.36)
Growth	681	(71.91)	589	(59.31)	673	(57.91)	1943	(62.64)
Total	947		993		1162		3102	

	Expected Frequencies			Deviations From Observed Frequencies			Chi-Square Components		
	A_1	A_2	A_3	A_1	A_2	A_3	A_1	A_2	A_3
Decline	354	371	434	-88	+33	+55	21.88	2.94	6.97
Growth	593	622	728	+88	-33	-55	13.06	1.75	4.16

X^2 (2)= 50.76 Sig.<.001

Population Change in the Urban Hierarchy 1950 - 1960 and Change in Number of Manufacturing Employees by Place of Residence 1950 - 1960.

Manufacturing Change	$a_1 + a_2$		a_3		a_4		Total	P
Decline	449	(37.48)	302	(34.72)	408	(39.46)	1159	(37.36)
Growth	749	(62.52)	568	(65.28)	626	(60.54)	1943	(62.64)
Total	1198		870		1034		3102	

	Expected Frequencies			Deviations From Observed Frequencies			Chi-Square Components		
	$a_1 + a_2$	a_3	a_4	$a_1 + a_2$	a_3	a_4	$a_1 + a_2$	a_3	a_4
Decline	448	325	386	+1	-23	+22	-	1.63	1.25
Growth	750	545	648	-1	+23	-22	-	.97	.75

X^2 (2)= 4.60 Sig.<.200

TABLE 4.2

Relative Accessibility (1950) and Change in Number of Manufacturing Employees by Place of Residence 1950 - 1960.

Manufacturing Change	B_1		B_2		B_3		Total	P
Decline	264	(34.47)	301	(29.37)	594	(45.31)	1159	(37.36)
Growth	502	(65.53)	724	(70.63)	717	(54.69)	1943	(62.64)
Total	766		1025		1311		3102	

	Expected Frequencies			Deviations From Observed Frequencies			Chi-Square Components		
	B_1	B_2	B_3	B_1	B_2	B_3	B_1	B_2	B_3
Decline	286	383	490	-22	-82	+104	1.69	17.56	22.07
Growth	480	642	821	+22	+82	-104	1.01	10.47	13.17

X^2 (2)= 65.97 Sig.< .001

Relative Accessibility (1960) and Change in Number of Manufacturing Employees by Place of Residence 1950 - 1960.

Manufacturing Change	B_1		B_2		B_3		Total	P
Decline	184	(38.82)	238	(27.39)	737	(41.90)	1159	(37.36)
Growth	290	(61.18)	631	(72.61)	1022	(58.10)	1943	(62.64)
Total	474		869		1759		3102	

	Expected Frequencies			Deviations From Observed Frequencies			Chi-Square Components		
	B_1	B_2	B_3	B_1	B_2	B_3	B_1	B_2	B_3
Decline	177	325	657	+7	-87	+80	.28	23.29	9.74
Growth	297	544	1102	-7	+87	-80	.16	13.91	5.81

X^2 (2)= 53.19 Sig. <.001

Change in Relative Accessibility and Change in Number of Manufacturing Employees by Place of Residence 1950 - 1960.

Manufacturing Change	B_1		B_2		Total	P
Decline	1044	(37.73)	115	(34.33)	1159	(37.36)
Growth	1723	(62.27)	220	(65.67)	1943	(62.64)
Total	2767		335		3102	

	Expected Frequencies		Deviations From Observed Frequencies		Chi-Square Components	
	B_1	B_2	B_1	B_2	B_1	B_2
Decline	1034	125	+10	-10	.10	.80
Growth	1733	210	-10	+10	.06	.48

X^2 (1)= 1.44 Sig.< .200

The dispersion deviation pattern remained the same in 1960 as in 1950. The increase in relative accessibility throughout the system did not change the extent of dispersion, meaning that many counties were doubly blessed by extension of the heartland as well as by dispersion. To the extent that a change can be detected in 1950 more than half of the chi-square deviation was due to group B_3. In 1960, 70 per cent of the deviation was due to group B_2, strengthening the evidence for dispersion.

Decentralization in Urban Fields

There was a "perfect" pattern of decentralization with respect to relative location of counties in urban fields in 1960. Table 4.3 reveals that as one goes from the group of counties closest to central cities toward counties at the periphery of the urban field, the proportion of counties experiencing growth in manufacturing employment by place of residence increased from 39.03 per cent to 71.53 per cent. There was extensive decentralization of place of residence during the 1950's, involving substantial increases in long distance commuting.

Hierarchical Diffusion and Dispersion

The interaction of hierarchical diffusion and dispersion of workers by place of residence is consistent and strong at all levels of the hierarchy, with the growth proportion being greatest in middle levels of national market access. In both 1950 and 1960, the interaction group A_1B_2 stands out in Table 4.4, with almost four out of five counties experiencing growth. This group alone contributed about 40 per cent to the overall value of the chi-square. Thus, hierarchical diffusion and dispersion were interacting throughout the system, but the interaction was particularly effective in medium-access counties within the nation. On the other hand, the growth of the manufacturing labor force was much less than expected in the high-access high-status FEA's.

TABLE 4.3

Relative Location in Urban Fields (1960) and Change in Number of Manufacturing Employees by Place of Residence 1952 - 1962.

Manufacturing Change

	C_1	C_2	C_3	C_4	Total P
Decline	187 (28.47)	404 (31.49)	254 (39.26)	314 (60.97)	1159 (37.36)
Growth	470 (71.53)	879 (68.51)	393 (60.74)	201 (39.03)	1943 (62.64)
Total	657	1283	647	515	3102

	Expected Frequencies				Deviations From Observed Frequencies				Chi-Square Components			
	C_1	C_2	C_3	C_4	C_1	C_2	C_3	C_4	C_1	C_2	C_3	C_4
Decline	246	479	242	192	-59	-75	+12	+122	14.15	11.74	.59	77.52
Growth	411	804	405	323	+59	+75	-12	-122	8.47	7.00	.36	46.08

X^2 (3)= 165.91 Sig.< .001

TABLE 4.4

Urban Hierarchy Interaction with Relative Accessibility (1950) and Change in Number of Manufacturing Employees by Place of Residence 1950 - 1960.

Interaction	Decline		Growth		Total
A_1B_1	98	(33.68)	193	(66.32)	291
A_1B_2	65	(19.23)	273	(80.77)	338
A_1B_3	103	(32.39)	215	(67.61)	318
A_2B_1	54	(48.22)	58	(51.78)	112
A_2B_2	75	(31.12)	166	(68.88)	241
A_2B_3	275	(42.97)	365	(57.03)	640
A_3B_1	32	(45.07)	39	(54.93)	71
A_3B_2	98	(33.80)	192	(66.20)	290
A_3B_3	359	(44.82)	442	(55.18)	801
Total	1159	(37.36)	1943	(62.64)	3102

	Expected Frequencies		Deviations From Observed Frequencies		Chi-Square Components	
	Decline	Growth	Decline	Growth	Decline	Growth
A_1B_1	109	182	-11	+11	1.11	.66
A_1B_2	126	212	-61	+61	29.53	17.55
A_1B_3	119	199	-16	+16	2.15	1.29
A_2B_1	42	70	+12	-12	3.43	2.06
A_2B_2	90	151	-15	+15	2.50	1.49
A_2B_3	239	401	+36	-36	5.42	3.23
A_3B_1	27	44	+ 5	- 5	.93	.57
A_3B_2	108	182	-10	+10	.93	.55
A_3B_3	299	502	+60	-60	12.04	7.17

X^2 (8)= 90.33 Sig. .001

Urban Hierarchy Interaction with Relative Accessibility (1960) and Change in Number of Manufacturing Employees by Place of Residence 1950 - 1960.

Interaction	Decline		Growth		Total
a_1b_1	144	(29.33)	347	(70.67)	491
a_1b_2	23	(20.00)	92	(80.00)	115
a_2b_1	269	(47.28)	300	(52.72)	569
a_2b_2	13	(56.53)	10	(43.47)	23
a_3b_1	287	(35.22)	528	(64.78)	815
a_3b_2	15	(27.28)	40	(72.72)	55
a_4b_1	344	(38.57)	548	(61.43)	892
a_4b_2	64	(45.07)	78	(54.93)	142
Total	1159	(37.36)	1943	(62.64)	3102

	Expected Frequencies		Deviations From Observed Frequencies		Chi-Square Components	
	Decline	Growth	Decline	Growth	Decline	Growth
a_1b_1	183	308	-39	+39	8.31	4.94
a_1b_2	43	72	-20	+20	9.30	5.56
a_2b_1	213	356	+56	-56	14.72	8.81
a_2b_2	9	14	+ 4	- 4	1.78	1.14
a_3b_1	304	511	-17	+17	.95	.57
a_3b_2	20	35	- 5	- 5	1.25	.71
a_4b_1	333	559	+11	-11	.36	.22
a_4b_2	54	88	+10	-10	1.85	1.14

X^2 (7)= 61.61 Sig.< .001

TABLE 4.4

Urban Hierarchy Interaction With Relative Accessibility (1950) and Change in Number of Manufacturing Employees by Place of Residence 1950-1960.

Interaction	Manufacturing Change				
	Decline		Growth		Total
A_1B_1	153	(31.16)	338	(68.84)	491
A_1B_2	77	(21.94)	274	(78.06)	351
A_1B_3	89	(39.38)	137	(60.72)	226
A_2B_1	73	(43.20)	96	(56.80)	169
A_2B_2	119	(32.34)	249	(67.66)	368
A_2B_3	226	(46.89)	256	(53.11)	482
A_3B_1	38	(35.85)	68	(64.15)	106
A_3B_2	105	(34.32)	201	(65.68)	306
A_3B_3	279	(46.27)	324	(53.77)	603
Total	1159	(37.36)	1943	(62.64)	3102

	Expected Frequencies		Deviations From Observed Frequencies		Chi-Square Components	
	Decline	Growth	Decline	Growth	Decline	Growth
A_1B_2	184	307	-31	+31	5.22	1.67
A_1B_2	131	220	-54	+54	22.26	13.25
A_1B_3	84	142	+ 5	- 5	.30	.18
A_2B_1	63	106	+10	-10	1.59	.94
A_2B_2	138	230	-19	+19	2.62	1.57
A_2B_3	180	302	+46	-46	11.76	7.01
A_3B_1	40	66	- 2	- 2	.10	.06
A_3B_2	114	192	- 9	- 9	.71	.42
A_3B_3	225	378	+54	-54	12.96	7.71

X^2 (8)= 90.33 Sig.< .001

The interaction between the change in FEA city population and change in county population potentials shows interesting deviations from expectations: the groups with no change or only average change in the population of their FEA centers - a_1b and a_3b - show more than a proportional number of counties experiencing growth in the number of manufacturing employees by place of residence. The reverse is true of those groups of counties whose FEA cities experienced either decline, a_2b or rapid growth a_4b. Thus, regardless of change in least relative accessibility, if the central city grew rapidly or declined, the chances of associated counties experiencing growth in manufacturing employment by place of residence was smaller than in counties associated with FEA cities that did not change or grew "normally." Counties associated with FEA cities that did not change were most likely to be located in peripheral areas of the country, the attractive areas for the location of industry in the 1950's. But these were areas that, at the same time, were in transformation from agriculture and mining to industrial occupations. It is much harder to identify the a_3b group. In any event, it seems that if an FEA city showed a dynamic character, be it decline or rapid growth, the conditions in counties associated with them were more ripe for change in all other manufacturing variables but the one discussed here. Change in manufacturing employment by place of residence seems to have been confined more to economic areas with stagnant urban centers than to areas whose urban centers were changing rapidly.

Hierarchical Diffusion and Decentralization

The interaction between hierarchical diffusion and decentralization confirms the outward spread of place of residence of industrial workers. Growth prospects were greater in the peripheral counties of smaller FEA's than in the central counties of high-status FEA's, and to compound the trends, groups A_1C_1 and A_1C_2 show positive deviations and groups A_2C_4 and A_3C_4 showed negative deviations. Clearly these four deviations represent the relative increase in industrial employment at the periphery of small urban centers, where the rural population was industrializing, and the relative decline of manufacturing employment in the established industrial cores of the country in favor of an increase in service occupations and extensive suburbanization of the manufacturing labor force.

The interaction of change in FEA city size 1950-1960 with relative location in urban fields in 1960 and the change in the number of manufacturing employees by place of residence is presented in the third part of Table 4.5. The impact of decentralization is quite systematic. The periphery of the urban fields, regardless of change in their FEA centers' populations, shows positive deviations. In fact the three groups $a_1+a_2C_1+C_2$, $a_3C_1+C_2$, and $a_4C_1+C_2$ together with a_3C_3 are the only groups to show positive deviations. The positive deviations become stronger in the periphery of urban centers that did not experience population growth. On the other hand, all other categories show negative deviations - especially where commuting levels in 1960 from county to FEA city were the highest.

Dispersion and Decentralization

The interaction between county population potential and relative location in an urban field is shown in Table 4.6. Starting with initial interaction, that of 1950, three points are worth explicit mention. First, about sixty per cent of the overall chi-square value is due to negative deviations of the groups of counties with the highest commuting levels to the FEA central cities - B_1C_4, B_2C_4, and B_3C_4. That is, regardless of initial level of relative accessibility, location in the core of the economic areas implied smaller chances of experiencing growth in manufacturing employment by place of residence. Second, the positive deviations are concentrated pri-

TABLE 4.5

Urban Hierarchy (1950) Interaction With Relative Location in Urban Field (1960) and Change in Number of Manufacturing Employees by Place of Residence 1950 - 1960.

Interaction	Decline		Growth		Total
A_1C_1	164	(27.11)	441	(72.89)	605
A_1C_2	58	(25.67)	168	(74.33)	226
A_1C_3	42	(32.56)	87	(67.44)	129
A_1C_4	55	(50.93)	53	(49.07)	108
A_2C_1	8	(47.06)	9	(52.94)	17
A_2C_2	166	(32.68)	342	(67.32)	508
A_2C_3	116	(42.03)	160	(57.97)	276
A_2C_4	128	(58.72)	90	(41.28)	218
A_3C_1	15	(42.86)	20	(57.14)	35
A_3C_2	180	(32.79)	369	(67.21)	549
A_3C_3	96	(39.67)	146	(60.33)	242
A_3C_4	131	(69.32)	58	(30.08)	189
Total	1159	(37.36)	1943	(62.64)	3102

	Expected Frequencies		Deviations From Observed Frequencies		Chi-Square Components	
	Decline	Growth	Decline	Growth	Decline	Growth
A_1C_1	226	379	-62	+62	17.01	10.14
A_1C_2	85	141	-27	+27	8.58	5.17
A_1C_3	48	81	- 6	+ 6	.75	.44
A_1C_4	40	68	+15	-15	5.62	3.31
A_2C_1	7	10	+ 1	- 1	.14	.10
A_2C_2	190	318	-24	+24	3.03	1.81
A_2C_3	103	173	+13	-13	1.64	.98
A_2C_4	81	137	+47	-47	27.27	16.12
A_3C_1	13	22	+ 2	- 2	.31	.18
A_3C_2	205	344	-25	+25	3.04	1.82
A_3C_3	90	152	+ 6	- 6	.40	.24
A_3C_4	71	118	+60	-60	50.70	30.51

X^2 (11)= 189.71 Sig. <.001

TABLE 4.5
(cont.)

Urban Hierarchy Interaction With Relative Location in Urban Fields (1960) and Change in Number of Manufacturing Employees by Place of Residence 1950 - 1960.

Interaction	Manufacturing Change: Decline		Growth		Total
A_1C_1	163	(27.04)	440	(72.96)	603
A_1C_2	38	(22.49)	131	(77.51)	169
A_1C_3	31	(31.64)	67	(68.36)	98
A_1C_4	34	(44.16)	43	(55.84)	77
A_2C_1	9	(50.00)	9	(50.00)	18
A_2C_2	161	(33.00)	327	(67.00)	488
A_2C_3	103	(38.29)	166	(61.71)	269
A_2C_4	131	(60.10)	87	(39.90)	218
A_3C_1	15	(41.67)	21	(58.33)	36
A_3C_2	205	(32.75)	421	(67.25)	626
A_3C_3	120	(42.86)	160	(57.14)	280
A_3C_4	149	(67.73)	71	(32.37)	220
Total	1159	(37.36)	1943	(62.64)	3102

	Expected Frequencies: Decline	Growth	Deviations Observed Frequencies: Decline	Growth	Chi-Square Components: Decline	Growth
A_1C_1	225	378	-62	+62	17.08	10.17
A_1C_2	63	106	-25	+25	9.92	5.90
A_1C_3	37	61	- 6	+ 6	.97	.59
A_1C_4	29	48	+ 5	- 5	.86	.52
A_2C_1	7	11	+ 2	- 2	.57	.36
A_2C_2	182	306	-21	+21	2.42	1.44
A_2C_3	101	168	+ 2	- 2	.04	.02
A_2C_4	81	137	+50	-50	30.86	18.25
A_3C_1	13	23	+ 2	- 2	.31	.17
A_3C_2	234	392	-29	+29	3.59	2.14
A_3C_3	105	175	+15	-15	2.14	1.29
A_3C_4	82	138	+67	-67	54.74	32.53

χ^2 (11) 196.88 Sig < .001

TABLE 4.5
(cont.)

Interaction of Change in F.E.A. City Size (1950-1960) with Relative Location in Urban Fields (1960) and Change in Number of Manufacturing Employees by Place of Residence 1950 - 1960.

Interaction	Manufacturing Change: Decline		Manufacturing Change: Growth		Total
$a_1+a_2C_1+C_2$	279	(30.36)	640	(69.64)	919
$a_1+a_2C_3$	76	(51.01)	73	(48.99)	149
$a_1+a_2C_4$	94	(72.31)	46	(27.69)	130
$a_3C_1+C_2$	122	(28.58)	305	(71.42)	427
a_3C_3	80	(32.13)	169	(67.87)	249
a_3C_4	100	(51.55)	94	(48.45)	194
$a_4C_1+C_2$	191	(32.16)	403	(67.84)	594
a_4C_3	97	(38.96)	152	(61.04)	249
a_4C_4	120	(62.86)	71	(37.14)	191
Total	1159	(37.36)	1943	(62.64)	3102

	Expected Frequencies		Deviations From Observed Frequencies		Chi-Square Components	
	Decline	Growth	Decline	Growth	Decline	Growth
$a_1+a_2C_1+C_2$	343	576	-64	+64	11.94	7.11
$a_1+a_2C_3$	56	93	+20	-20	7.14	4.30
$a_1+a_2C_4$	49	81	+45	-45	41.33	25.00
$a_3C_1+C_2$	160	267	-38	+38	9.02	5.41
a_3C_3	93	156	-13	+13	1.82	1.08
a_3C_4	72	122	+28	-28	10.89	6.43
$a_4C_1+C_2$	222	372	-31	+31	4.33	2.58
a_4C_3	93	156	+ 4	- 4	.17	.10
a_4C_4	71	120	+49	-49	33.82	20.01

X^2 (8)= 192.48 Sig. $<.001$

TABLE 4.6

Relative Accessibility (1950) Interaction With Relative Location in Urban Field (1960) and Change in Number of Manufacturing Employees by Place of Residence 1950 - 1960

Interaction	Manufacturing Change				
	Decline		Growth		Total
B_1C_1	93	(27.77)	242	(72.23)	335
B_1C_2	73	(29.20)	177	(70.80)	250
B_1C_3	32	(38.10)	52	(61.90)	84
B_1C_4	66	(68.05)	31	(31.95)	97
B_2C_1	37	(20.67)	142	(79.33)	179
B_2C_2	115	(23.91)	366	(76.09)	481
B_2C_3	71	(31.99)	151	(68.01)	222
B_2C_4	78	(54.55)	65	(45.45)	143
B_3C_1	57	(39.86)	86	(60.14)	143
B_3C_2	216	(39.13)	336	(60.87)	552
B_3C_3	150	(43.99)	191	(56.01)	341
B_3C_4	171	(62.19)	104	(37.81)	275
Total	1159	(37.36)	1943	(62.64)	3102

	Expected Frequencies		Deviations From Observed Frequencies		Chi-Square Components	
	Decline	Growth	Decline	Growth	Decline	Growth
B_1C_1	125	210	-32	+32	8.19	4.87
B_1C_2	93	157	-20	+20	4.30	2.55
B_1C_3	32	52	---	---	---	---
B_1C_4	36	61	+30	-30	25.00	14.75
B_2C_1	67	112	-30	+30	13.43	8.04
B_2C_2	180	301	-65	+65	23.47	14.04
B_2C_3	83	139	-12	+12	1.74	1.04
B_2C_4	53	90	+25	-25	11.79	6.94
B_3C_1	53	90	+ 4	- 4	.30	.18
B_3C_2	206	346	+10	-10	.49	.29
B_3C_3	128	213	+22	-22	3.78	2.27
B_3C_4	103	172	+68	-68	44.89	26.88

TABLE 4.6
(cont.)

Relative Accessibility Interaction with Relative Location in Urban Field (1960) and Change in Number of Manufacturing Employees by Place of Residence 1950 - 1960

Interaction	Manufacturing Change				
	Decline		Growth		Total
B_1C_1	76	(32.48)	158	(67.52)	234
B_1C_2	50	(35.46)	91	(64.54)	141
B_1C_3	22	(46.81)	25	(53.19)	47
B_1C_4	36	(69.23)	16	(30.77)	52
B_2C_1	36	(18.19)	162	(81.81)	198
B_2C_2	88	(21.84)	315	(78.16)	403
B_2C_3	47	(31.76)	101	(68.24)	148
B_2C_4	67	(55.84)	53	(44.16)	120
B_3C_1	75	(33.34)	150	(66.66)	225
B_3C_2	266	(36.00)	473	(64.00)	739
B_3C_3	185	(40.93)	267	(59.07)	452
B_3C_4	211	(61.52)	132	(38.48)	343
Total	1159	(37.36)	1943	(62.64)	3102

	Expected Frequencies		Deviations From Observed Frequencies		Chi-Square Components	
	Decline	Growth	Decline	Growth	Decline	Growth
B_1C_1	87	147	-11	+11	1.39	.82
B_1C_2	53	88	- 3	+ 3	.17	.10
B_1C_3	18	29	+ 4	- 4	.89	.55
B_1C_4	19	33	+17	-17	15.21	8.76
B_2C_1	74	124	-38	+38	19.51	11.64
B_2C_2	151	252	-63	+63	26.28	15.75
B_2C_3	55	93	- 8	+ 8	1.16	.69
B_2C_4	45	75	+22	-22	10.76	6.45
B_3C_1	84	141	- 9	+ 9	.96	.57
B_3C_2	276	463	-10	+10	.36	.22
B_3C_3	169	283	+16	-16	1.51	.90
B_3C_4	128	215	+83	-83	53.82	32.04

X^2 (11)= 210.51 Sig. < .001

TABLE 4.6
(cont.)

Interaction of Change in Relative Accessibility (1950 - 1960) With Relative Location in Urban Fields (1964) and Change in Number of Manufacturing Employees by Place of Residence 1950 - 1960.

Interaction	Manufacturing Change				
	Decline		Growth		Total
b_1C_1	162	(30.28)	373	(69.72)	535
b_1C_2	387	(32.31)	811	(67.69)	1198
b_1C_3	240	(39.80)	363	(60.20)	603
b_1C_4	255	(59.17)	176	(40.83)	431
b_2C_1	26	(21.32)	96	(78.68)	122
b_2C_2	17	(20.00)	68	(80.00)	85
b_2C_3	13	(29.55)	31	(71.45)	44
b_2C_4	59	(70.24)	25	(29.76)	84
Total	1159	(37.36)	1943	(62.64)	3102

	Expected Frequencies		Deviations From Observed Frequencies		Chi-Square Components	
	Decline	Growth	Decline	Growth	Decline	Growth
b_1C_1	200	335	-38	+38	7.22	4.31
b_1C_2	448	750	-61	+61	8.31	4.96
b_1C_3	225	378	+15	-15	1.00	.60
b_1C_4	161	270	+94	-94	54.88	32.73
b_2C_1	46	76	-20	+20	8.70	5.63
b_2C_2	32	53	-15	+15	7.03	4.24
b_2C_3	16	28	- 3	+ 3	.56	.32
b_2C_4	31	53	+28	-28	25.29	14.79

X^2 (7)= 180.57 Sig. $<$ .001

marily in two interaction groups: B_2C_1 and B_2C_2, in fact about 27 per cent of the variations came from these two groups. This is to say that the greatest chances for experiencing growth are in counties that were located on the periphery of urban fields that simultaneously possessed medium levels of national market access in 1950. Third, groups of counties with high levels of national market access, regardless of their relative location in urban fields, all show negative deviations in the proportion of growing counties.

When the change in manufacturing employment is viewed from the vantage of relative location in 1960, the deviations are similar to that of the 1950. The BC_4 groups again show systematically high negative deviations. The positive deviations are concentrated in the B_2C_1 and B_2C_2 groups. Together, these five groups (BC_4, B_2C_1 and B_2C_2) are responsible for 86 per cent of the chi-square value.

These interaction deviations indicate that during the 1950's there was systematic interaction of dispersion and decentralization: counties that were located on the periphery of urban fields and that had low or medium population potential were more likely to experience growth in manufacturing employment by place of residence than counties that were located closer to the centers of their economic areas and possessed or achieved high levels of relative accessibility.

Factors Associated with Growth-Rate Differentials

Tables 4.7-4.10 enable us to turn to the variance-covariance part of our analysis of change in the residential location of manufacturing workers. Table 4.7 reveals that the initial relative location of counties in 1950 was associated with the rate of change of counties' manufacturing employment by place of residence. However, the three basic components of relative location exerted independent influences on the rate of change, because most interactions between pairs of relative location components were not statistically significant. Influence of the three main effects is evident throughout Table 4.7. In the first part of the table, where the order of effects is arbitrarily specified, only one interaction effect, AB, is statistically significant, and its F-ratio is very low relative to the main effects. In the second part of the table, where each effect appears alone, however, all effects are statistically significant - but this is due to the three main effects rather than the interaction effects. Specifying the order of effects using significant effects only results in the particular combination presented in the third part of the table. Apparently, initial relative accessibility is the most powerful effect on change in manufacturing employment by place of residence. When this effect is held constant, the second effect is that of a county's association with an urban center of a given level in the national urban hierarchy. When both the above effects are controlled for, the location of counties in their urban field is still important. The fourth effect to influence manufacturing change is the interaction of hierarchical status and relative accessibility, AB.

Least-squares estimates of the significant subgroup growth-rate deviations from the national average (presented in the third part of the table) clarify the relationships. Not only did the proportion of counties growing increase as a consequence of diffusion, dispersion and decentralization, but at the extremes of relative location the growth rates far exceeded the national average, while in the cores of high-status urban regions within the heartland, growth rates were significantly lower, and in the most central case, there was actual decline of the group as a whole.

The same conclusions can be drawn from the analysis with respect to relative location in 1960 (Table 4.8). All that was said with

TABLE 4.7

Analysis of Variance 1950

Effect	Mean Square	Degrees of Freedom	F. Ratio	Significance
A	11750.24	2 ,3066	44.32	<.0001
B	5315.06	2 ,3066	20.05	<.0001
C	3278.43	3 ,3066	12.37	<.0001
AB	886.73	4 ,3066	3.34	<.0097
AC	94.81	6 ,3066	.36	<.9063
BC	519.05	6 ,3066	1.96	<.0676
ABC	168.17	12 ,3066	.63	<.8147

ANOVA: Each Effect Alone

Effect	Mean Square	Degrees of Freedom	F. Ratio	Significance
A	11750.24	2 ,3066	44.32	<.0001
B	11852.21	2 ,3066	44.70	<.0001
C	7768.83	3 ,3066	29.30	<.0001
AB	1453.42	4 ,3066	5.41	<.0003
AC	1353.82	6 ,3066	5.11	<.0001
BC	1249.02	6 ,3066	4.71	<.0001
ABC	1611.56	12 ,3066	6.08	<.0001

ANOVA: Significant Effects Only

Effect	Mean Square	Degrees of Freedom	F. Ratio	Significance
B	11852.51	2 ,3066	44.70	<.0001
A	5213.09	2 ,3066	19.66	<.0001
C	3278.44	3 ,3066	12.37	<.0001
AB	886.73	4 ,3066	3.34	<.0097

Least Square Estimates of Effects

General Mean	+8.28
B_1	+4.08
B_2	+3.33
A_1	+4.41
A_2	+1.54
C_1	+4.57
C_2	+5.15
C_3	+3.20
A_1B_1	+4.84
A_1B_2	+2.61
A_2B_1	-1.61
A_2B_2	-2.24

TABLE 4.8

Analysis of Variance (1960)

Effect	Mean Square	Degrees of Freedom	F. Ratio	Significance
A	12878.47	2 ,3066	48.69	<.0001
B	6014.57	2 ,3066	22.74	<.0001
C	3070.30	3 ,3066	11.61	<.0001
AB	882.74	4 ,3066	3.34	<.0098
AC	156.46	6 ,3066	.59	<.7383
BC	302.42	6 ,3066	1.14	<.3314
ABC	164.89	12 ,3066	.62	<.8245

ANOVA: Each Effect Alone

Effect	Mean Square	Degrees of Freedom	F. Ratio	Significance
A	12878.47	2 ,3066	48.69	<.0001
B	11106.47	2 ,3066	42.00	<.0001
C	7769.31	3 ,3066	29.38	<.0001
AB	861.12	4 ,3066	3.27	<.0110
AC	1459.73	6 ,3066	5.52	<.0001
BC	755.28	6 ,3066	2.86	<.0089
ABC	1075.52	12 ,3006	4.07	<.0001

ANOVA: Significant Effects Only

	Effect	Mean Square	Degrees of Freedom	Significance
1.	A	12878.47	2 ,3066	<.0001
	B	6014.57	3 ,3066	<.0001
	C	3070.30	3 ,3066	<.0001
	AB	882.74	4 ,3066	<.0098
2.	A	12878.47	2 ,3066	<.0001
	B	6014.57	2 ,3066	<.0001
	C	3070.30	3 ,3066	<.0001
	BC	537.80	6 ,3066	<.0575

Least Square Estimates of Effects

Alternative 2:

General Mean	+8.94
A_1	+5.50
A_2	+2.07
B_1	+2.39
B_2	+4.09
C_1	+5.24
C_2	+5.55
C_3	+4.20
B_1C_1	+4.71
B_1C_2	+7.41
B_1C_3	+ .82
B_2C_1	+4.27
B_2C_2	+2.82
B_2C_3	+3.12

Alternative 1: N.A.

respect to the initial conditions pertains also to the final ones. In addition, changes in the place of residence of manufacturing workers are closely related (Table 4.9). Four components of change are highly significant, and two alternative orderings of the significant factors appear equally acceptable. Alternative 2 assigns the greatest effect to change in relative accessibility and the relative location of counties in their urban field. Alternative 1 reduces the overwhelming effect of b and C and puts the hierarchical effect as the most important one instead of the smaller role assigned to it in alternative 2. The direction for the subeffect C is as found earlier and as expected, the effect of rapid change in population potential (b_2) is positive while that of b_1 ("normal" population potential) is negative. The a_i subeffects indicate a familiar pattern: where the FEA city population grew most rapidly during the 1950's, the increase in manufacturing employment by place of residence tended to be relatively less than in counties associated with FEA cities that grew normally or did not change. That is, a dynamic FEA center during the 1950's implied <u>less</u> growth of manufacturing employment by place of residence. The reverse was true for "stagnant" FEA centers (a_1--no change). This probably indicates that the mobility of the labor force increased much faster than the ability of manufacturing plants to locate everywhere.

Combining the initial relative locations and locational change (Table 4.10), three sources of variation in the manufacturing employment residence variable stand out: Bb, Aa, and C. Three possible combinations of these three effects are all statistically significant. The conventional alternative is alternative No. 3, which shows the three effects in order of their statistical power, but which alternative one prefers is a matter of judgement. It seems that with respect to the present variable, the third alternative should be pre-

TABLE 4.9

Analysis of Variance (1950-1960)

Effect	Mean Square	Degrees Of Freedom	F. Ratio	Significance
a	5618.82	3 ,3074	21.08	<.0001
b	5291.04	1 ,3074	19.86	<.0001
C	3956.50	3 ,3074	14.85	<.0001
ab	2196.55	3 ,3074	8.24	<.0001
aC	158.30	9 ,3074	.59	<.8033
bC	899.93	3 ,3074	3.38	<.0175
abC	N.A		N.A.	< N.A.

ANOVA: Each Effect Alone

Effect	Mean Square	Degrees Of Freedom	F. Ratio	Significance
a	5618.82	3 ,3074	21.08	<.0001
b	7802.49	1 ,3074	29.28	<.0001
C	7768.62	3 ,3074	29.15	<.0001
ab	706.62	3 ,3074	2.65	<.0467
aC	3107.53	9 ,3074	11.66	<.0001
bC	792.32	3 ,3074	2.97	<.0303
abC	158.10	9 ,3074	.59	<.8042

ANOVA: Significant Effects Only

	Effect	Mean Square	Degrees Of Freedom	F. Ratio	Significance
1.	a	5618.82	3 ,3074	21.08	<.0001
	b	5291.04	1 ,3074	19.86	<.0001
	C	3956.50	3 ,3074	14.85	<.0001
	ab	2196.55	3 ,3074	8.24	<.0001
2.	b	7802.49	1 ,3074	29.28	<.0001
	C	7560.42	3 ,3074	28.37	<.0001
	bC	1343.36	3 ,3074	5.04	<.0018
	a	1078.84	3 ,3074	4.05	<.0070

(A number of other alternatives is available.)

Least Square Estimates of Effects

General Mean	+10.37
b_1	- 4.73
C_1	+ 6.86
C_2	+ 8.54
C_3	+ 6.04
b_1C_1	- 8.98
b_1C_2	- 7.98
b_1C_3	- 7.08
a_1	+ 5.50
a_2	- .54
a_3	+ 1.48

TABLE 4.10

Analysis of Variance--1950 and 1950-1960

Effect	Mean Square	Degrees of Freedom	F. Ratio	Significance
Aa	2715.15	6 ,2959	10.45	<.0001
Bb	3630.76	2 ,3959	13.98	<.0001
C	4827.92	3 ,2959	18.59	<.0001
AaBb	199.82	12 ,2959	.77	<.6834
AaC	352.14	18 ,2959	1.36	<.1431
BbC	270.48	6 ,2959	1.04	<.3953

ANOVA: Each Effect Alone

Effect	Mean Square	Degrees of Freedom	F. Ratio	Significance
Aa	2715.15	6 ,2959	10.45	<.0001
Bb	5266.63	2 ,2959	20.27	<.0001
C	7769.23	3 ,2959	29.91	<.0001
AaBb	976.60	12 ,3959	3.76	<.0001
AaC	1463.34	18 ,2959	5.63	<.0001
BbC	606.99	6 ,2959	2.34	<.0296

ANOVA: Significant Effects Only

Effect	Mean Square	Degrees of Freedom	F. Ratio	Significance
Bb	5266.62	2 ,2959	20.27	<.0001
Aa	2169.86	6 ,2959	8.35	<.0001
C	4827.92	3 ,2959	18.59	<.0001
Aa	2715.15	6 ,2959	10.45	<.0001
Bb	3630.76	2 ,2959	12.98	<.0001
C	4827.92	3 ,2959	18.59	<.0001
C	7769.23	3 ,2959	29.91	<.0001
Bb	3294.61	2 ,2959	12.68	<.0001
Aa	1356.55	6 ,2969	5.22	<.0001

Least Square Estimates of Effects

General Mean	+ 9.51
C_1	+10.32
C_2	+ 4.97
C_3	+ 2.91
B_1b_1	+ 2.51
B_1b_2	+ 5.76
A_1a_1	- 9.94
A_1a_2	- .23
A_1a_3	+ 3.49
A_2a_1	- 8.43
A_2a_2	+ 6.74
A_2a_3	+ 2.55

ferred. Change in manufacturing employment by place of residence seems to be consistently associated with relative location of counties in their urban field . . . with what happened in C_1, C_2 and C_3 areas that were primarily rural in character during the 1950's and were gradually shifting toward greater industrialization of the labor force. First, the surplus labor force and that part that sought additional income moved toward the jobs, and only later was there an increase in jobs moving to rural areas. Decentralization of manufacturing employment to the periphery of urban fields was dependent on initial relative accessibility (1950) simultaneously with the change in this accessibility, and with hierarchical status and the nature of change of this status.

CHANGES IN THE DISTRIBUTION OF EMPLOYMENT

The change in the number of manufacturing employees in a given county -- the location of industrial jobs -- is not necessarily congruent with the change in the number of employees in manufacturing who reside in the same county. Thus, increased employment by place of work implies that a county is industrializing but the workers, at the extreme, could all reside in a different county. The distinction is essential for the interpretation of results from the chi-square analysis (Table 4.11 ff). One might expect the change in the total number of manufacturing establishments, discussed already, to conform to the change in the number of manufacturing employees but this is not so. A number of factors serve to differentiate the two, and so the employment variable deserves specific attention. We can find counties that experienced growth in both number of establishments and employment as well as counties that experienced decline in both. Thus, a full relative location analysis of the variable should be worthwhile.

Hierarchical Diffusion

The average change per county in the number of manufacturing employees working in it was 53.4 per cent for the 1952-1962 decade. Did the changes vary with the hierarchical level of FEA centers? Table 4.11 shows that there were indeed statistically significant associations. The population size of FEA cities in both 1950 and in 1960, and the change in their population between 1950 and 1960, are all associated with the change in manufacturing employment by place of work. Further, the direction of deviations from expectations follows a familiar pattern. More counties than expected associated with small FEA cities experienced decline. In this sense, the results are no different from the ones achieved with respect to each the total establishments variable. However, the chi-square components are smaller and concentrated explicitly in the small FEA city category. At the same time it is clear that hierarchical diffusion of change is pervasive.

Dispersion

Dispersion of employment change by place of work also is evident (Table 4.12). Counties in all categories of relative accessibility experienced growth. In 1950 the negative effect of low population potential is evident, and so is the positive effect of high relative accessibility. This could be a result of the improvement in accessibility throughout the system in the 1960's, as well as the relative increase of the continuing attractiveness of counties that experienced growth in manufacturing employment by place of work had high relative accessibility by 1960. Rapid growth in accessibility significantly increases the likelihood of growth in manufacturing employment by place of work. Counties that increased their population potential by less than twenty per cent during the 1950's had a 58 per cent chance the manufacturing jobs available in them would in-

TABLE 4.11

Urban Hierarchy (1950) and Change in Number of Manufacturing Employees by Place of Work 1952 - 1962.

Manufacturing Change	A_1	A_2	A_3	Total	P
Decline	510 (47.75)	389 (38.17)	375 (36.94)	1274	(41.07)
Growth	558 (52.25)	630 (61.83)	640 (63.06)	1828	(58.93)
Total	1068	1019	1015	3102	

	Expected Frequencies			Deviations From Observed Frequencies			Chi-Square Components		
	A_1	A_2	A_3	A_1	A_2	A_3	A_1	A_2	A_3
Decline	439	418	417	+71	-29	-42	11.48	2.01	4.23
Growth	629	601	598	-71	+29	+42	8.01	1.40	2.95

X^2 (2)= 30.08 Sig.< .001

Urban Hierarchy (1960) and Change in Number of Manufacturing Employees by Place of Work 1952 - 1962.

Manufacturing Change	A_1	A_2	A_3	Total	P
Decline	446 (49.21)	370 (37.26)	438 (37.69)	1274	(41.07)
Growth	481 (50.79)	623 (62.74)	724 (62.31)	1828	(58.93)
Total	947	993	1162	3102	

	Expected Frequencies			Deviations From Observed Frequencies			Chi-Square Components		
	A_1	A_2	A_3	A_1	A_2	A_3	A_1	A_2	A_3
Decline	389	408	477	+77	-38	-39	15.24	3.54	3.19
Growth	558	585	685	-77	+38	+39	10.62	2.47	2.22

X^2 (2)= 37.28 Sig.< .001

Population Change in the Urban Hierarchy 1950 - 1960 and Change in Number of Manufacturing Employees by Place of Work 1952 - 1962.

Manufacturing Change	a_1+a_2	a_3	a_4	Total	P
Decline	548	342	384	1247	(41.07)
Growth	650	528	650	1828	(58.93)

	Expected Frequencies			Deviations From Observed Frequencies			Chi-Square Components		
	a_1+a_2	a_3	a_4	a_1+a_2	a_3	a_4	a_1+a_2	a_3	a_4
Decline	492	357	425	+56	-15	-41	6.37	.63	3.96
Growth	706	513	609	-56	+15	+41	4.44	.44	2.76

X^2 (2)= 18.60 Sig.< .001

TABLE 4.12

Relative Accessibility (1950) and Change in Number of Manufacturing Employees by Place of Work 1952 - 1962.

Manufacturing Change	B_1		B_2		B_3		Total	P
Decline	385	(50.26)	415	(40.49)	474	(36.16)	1274	(41.07)
Growth	381	(49.74)	610	(59.51)	837	(63.84)	1828	(58.93)
Total	766		1025		1311		3102	

	Expected Frequencies			Deviations From Observed Frequencies			Chi-Square Components		
	B_1	B_2	B_3	B_1	B_2	B_3	B_1	B_2	B_3
Decline	315	421	538	+70	-6	-64	15.56	.08	7.61
Growth	451	604	773	-70	+6	+64	10.86	.06	5.30

X^2 (2)= 39.47 Sig. < .001

Relative Accessibility (1960) and Change in Number of Manufacturing Employees by Place of Work 1952 - 1962.

Manufacturing Change	B_1		B_2		B_3		Total	P
Decline	234	(49.37)	400	(46.03)	640	(36.38)	1274	(41.07)
Growth	240	(50.63)	469		1119	(63.62)	1828	(58.93)
Total	474		869		1759		3102	

	Expected Frequencies			Deviations From Observed Frequencies			Chi-Square Components		
	B_1	B_2	B_3	B_1	B_2	B_3	B_1	B_2	B_3
Decline	195	357	722	+39	+43	-82	7.80	5.18	9.31
Growth	279	512	1037	-39	-43	+82	5.45	3.61	6.48

X^2 (2)= 37.83 Sig.<.001

Change in Relative Accessibility (1950 - 1960) and Change in Number of Manufacturing Employees by Place of Work 1952 - 1962.

Manufacturing Change	b_1		b_2		Total	P
Decline	1167	(42.18)	107	(31.94)	1274	(41.07)
Growth	1600	(57.82)	228	(68.06)	1828	(58.93)
Total	2767		335		3102	

	Expected Frequencies		Deviations From Observed Frequencies		Chi-Square Components	
	b_1	b_2	b_1	b_2	b_1	b_2
Decline	1136	138	+31	-31	.85	6.96
Growth	1631	197	-31	+31	.59	4.98

X^2 (1)= 13.28 Sig.< .001

crease. In counties that experienced more than twenty per cent increase in relative accessibility, the probability is 68 per cent.

Decentralization in Urban Fields

Decentralization within urban fields also is apparent (Table 4.13). In the no-commuting zone (C_1) one out of two counties increased its employment in manufacturing. This fraction rises to 60 per cent in counties in the most accessible commuting zone (C_4). The two areas that showed significant deviations from the expected proportions are C_1, with negative deviations, and C_4 with positive deviations. The pattern of deviation is the same as in the cases of change in the number of manufacturing establishments.

Hierarchical Diffusion and Dispersion

In addition to the simultaneous hierarchical diffusion and dispersion that is shown in Table 4.14, few systematic subgroup deviations are apparent. For many A_iB_i groups the chi-square components are very low. In contrast to the high positive effect of B_3 on A and of A_3 on B that was evident in the change in number of establishments, 67 per cent of the overall chi-square value arises from the negative deviation of the A_1B_1 group. Otherwise, counties behave exactly as expected on the basis of the joint probability derived from the margins. Being in the periphery of the country in 1950 or in 1960 contributed to lagging growth in manufacturing employment.

Not surprisingly, then, the interaction between relative accessibility change and FEA city-size change is associated with change in manufacturing employment by place of work. The positive effect of rapid changes in population potential and/or in FEA city populations is clearly evident. Also, the negative deviations of a_1b_1 are an indication that change in relative accessibility alone is not sufficient to stimulate growth. To create growth, both improvements in access and a rapidly growing urban center are apparently needed.

TABLE 4.13

Relative Location in Urban Fields (1960) and Change in Number of Manufacturing Employees by Place of Work 1952 - 1962.

Manufacturing Change

	C_1	C_2	C_3	C_4	Total	P
Decline	324 (49.32)	521 (40.61)	258 (39.88)	171 (33.20)	1274	(41.07)
Growth	333 (50.68)	762 (59.39)	389 (60.12)	344 (66.80)	1828	(58.93)
Total	657	1283	647	515	3102	

	Expected Frequencies				Deviations From Observed Frequencies				Chi-Square Components			
	C_1	C_2	C_3	C_4	C_1	C_2	C_3	C_4	C_1	C_2	C_3	C_4
Decline	270	527	266	211	+54	-6	-8	-40	10.80	.07	.24	7.58
Growth	387	756	381	304	-54	+6	+8	+40	7.53	.05	.17	5.26

X^2 (3)= 31.70 Sig. .001

TABLE 4.14

Urban Hierarchy Interaction With Relative Accessibility (1950) and Change in Number of Manufacturing Employees by Place of Work 1952 - 1962.

Interaction	Manufacturing Change				
	Decline		Growth		Total
A_1B_1	283	(57.64)	208	(42.36)	491
A_1B_2	160	(45.58)	191	(54.42)	351
A_1B_3	67	(29.65)	159	(70.05)	226
A_2B_1	66	(39.05)	103	(60.95)	169
A_2B_2	138	(37.50)	230	(62.50)	368
A_2B_3	185	(38.38)	297	(61.62)	482
A_3B_1	36	(33.96)	70	(66.04)	106
A_3B_2	117	(38.24)	189	(61.76)	306
A_3B_3	222	(36.82)	381	(63.18)	603
Total	1274	(40.07)	1828	(59.93)	3102

	Expected Frequencies		Deviations From Observed Frequencies		Chi-Square Components	
	Decline	Growth	Decline	Growth	Decline	Growth
A_1B_1	202	289	+81	-81	32.48	22.70
A_1B_2	144	207	+16	-16	1.78	1.24
A_1B_3	93	133	-26	+26	7.27	5.08
A_2B_1	69	100	- 3	+ 3	.13	.09
A_2B_2	151	217	-13	+13	1.12	.78
A_2B_3	198	284	-13	+13	.85	.60
A_3B_1	43	63	- 7	+ 7	1.14	.78
A_3B_2	126	180	- 9	+ 9	.64	.45
A_3B_3	248	355	-26	+26	2.73	1.90

X^2 (8)= 81.76 Sig. < .001

Urban Hierarchy Interaction With Relative Accessibility (1950) and Change in Number of Manufacturing Employees by Place of Work 1952 - 1962.

Interaction	Manufacturing Change				
	Decline		Growth		Total
A_1B_1	164	(56.36)	127	(43.64)	291
A_1B_2	193	(57.10)	145	(42.90)	338
A_1B_3	109	(34.28)	211	(65.72)	318
A_2B_1	43	(38.39)	69	(61.61)	112
A_2B_2	88	(36.51)	153	(63.49)	241
A_2B_3	239	(37.34)	401	(62.66)	640
A_3B_1	27	(38.03)	34	(61.97)	71
A_3B_2	119	(41.03)	171	(58.97)	290
A_3B_3	292	(36.45)	509	(63.55)	801
	1274	(41.07)	1828	(58.93)	3102

	Expected Frequencies		Deviations From Observed Frequencies		Chi-Square Components	
	Decline	Growth	Decline	Growth	Decline	Growth
A_1B_1	188	103	+44	-44	16.13	18.80
A_1B_2	218	120	+54	-54	20.98	24.30
A_1B_3	205	113	-22	+22	3.69	4.28
A_2B_1	72	40	- 3	+ 3	.20	.22
A_2B_2	155	86	-11	+11	1.22	1.41
A_2B_3	413	227	-24	+24	2.19	2.54
A_3B_1	46	25	- 2	+ 2	.14	.16
A_3B_2	187	113	-	-	-	-
A_3B_3	517	284	-36	+36	3.95	4.56

X^2 (8)= 104.77 Sig. < .001

TABLE 4.14
(cont.)

Interaction of Change in F.E.A. City Size With Change in Relative Accessibility (1950 - 1960) and Change in Number of Manufacturing Employees by Place of Work 1952 - 1962

Interaction	Manufacturing Change				
	Decline		Growth		Total
a_1b_1	246	(50.10)	245	(49.90)	491
a_1b_2	61	(53.04)	54	(46.96)	115
a_2b_1	238	(41.83)	331	(58.17)	569
a_2b_2	3	(13.04)	20	(86.96)	23
a_3b_1	329	(40.37)	486	(59.63)	815
a_3b_2	13	(23.64)	42	(76.36)	55
a_4b_1	354	(39.69)	538	(60.39)	829
a_4b_2	30	(21.13)	112	(78.87)	142
Total	1274	(41.07)	1828	(58.93)	3102

	Expected Frequencies		Deviations From Observed Frequencies		Chi-Square Components	
	Decline	Growth	Decline	Growth	Decline	Growth
a_1b_1	202	298	+44	-44	9.58	6.50
a_1b_2	47	68	+14	-14	4.17	2.88
a_2b_1	234	335	+ 4	- 4	.07	.05
a_2b_2	9	14	- 6	+ 6	4.00	2.57
a_3b_1	335	480	- 6	+ 6	.11	.08
a_3b_2	23	32	-10	+10	4.35	3.12
a_4b_1	366	526	-12	+12	.39	.27
a_4b_2	58	84	-28	+28	13.52	9.33

X^2 (6)= 60.99 Sig. < .001

Hierarchical Diffusion and Decentralization

There appears, on the other hand, to be little relationship between hierarchical status and commuting levels (Table 4.15). What was evident in Table 4.14 with respect to hierarchical diffusion and dispersion repeats itself. The negative effect of peripheral locatio (areas A_1C_1 and A_1C_2) is clear, as is the strong positive effect of high levels of commuting (C_4). One suspects that the relative location of counties has a more selective effect on the change in the number of manufacturing plants than it has on employment change by place of work, however, perhaps because many more factors unrelated to relative location influence the change in employment than the change in number of establishments.

The change in relative location to the 1960 levels does change the relationships that were evident with respect to the 1950 conditions. With respect to the interaction between change in the FEA city population and the location of counties in their urban field, most of the positive variation comes from group a_4C_4, the group of counties that was associated with rapidly growing FEA cities and with substantial daily commuting to such cities. Indeed one suspects that this highly positive deviation is a reflection of rapid suburbanization. It is probably in the central county of an economic area and those immediately contiguous areas that increases in employment in manufacturing by place of work are the most likely. In fact four out of five counties in A_4C_4 group experienced growth.

Dispersion and Decentralization

As with all other interactions, so it is in the case of the simultaneous effect of dispersion and decentralization: high negative deviations in the marginal areas B_1C_1 and B_1C_2 (Table 4.16), 56.6 per cent of overall chi-square value coming from the negative deviations of these two groups. The positive effect of C_4 inter-

TABLE 4.15

Urban Hierarchy (1950) Interaction with Relative Location in Urban Field (1960) and Change in Number of Manufacturing Employees by Place of Work (1952 - 1962).

Interaction	Manufacturing Change: Decline		Growth		Total
A_1C_1	308	(50.91)	297	(49.19)	605
A_1C_2	117	(51.77)	109	(48.23)	226
A_1C_3	53	(41.08)	76	(58.92)	129
A_1C_4	22	(29.63)	76	(70.37)	108
A_2C_1	5	(29.41)	12	(70. 9)	17
A_2C_2	201	(39.57)	377	(60.43)	508
A_2C_3	108	(39.13)	168	(60.87)	276
A_2C_4	75	(34.40)	143	(5.60)	218
A_3C_1	11	(31.43)	24	(68.57)	35
A_3C_2	203	(36.98)	346	(63.02)	549
A_3C_3	97	(40.08)	145	(59.92)	242
A_3C_4	64	(33.86)	125	(66.14)	189
Total	1274	(41.07)	1828	(58.93)	3102

	Expected Frequencies		Deviations From Observed Frequencies		Chi-Square Components	
	Decline	Growth	Decline	Growth	Decline	Growth
A_1C_1	249	356	+59	-59	14.00	9.78
A_1C_2	93	133	+24	-24	6.19	4.33
A_1C_3	53	76	---	---	----	---
A_1C_4	44	64	-12	+12	3.27	2.25
A_2C_1	7	10	- 2	+ 2	.57	.40
A_2C_2	209	299	- 8	+ 8	.31	.21
A_2C_3	113	163	- 5	+ 5	.22	.15
A_2C_4	90	128	-15	+15	2.50	1.76
A_3C_1	14	21	- 3	+ 3	.64	.33
A_3C_2	225	324	-22	+22	2.15	1.49
A_3C_3	99	143	- 2	+ 2	.04	.03
A_3C_4	78	111	-14	+14	2.51	1.77

X^2 (11)= 54.90 Sig.< .001

TABLE 4.15

(cont.)

Urban Hierarchy interaction with Relative Location in Urban Fields (1960) and Change in Number of Manufacturing Employees by Place of Work 1952 - 1962.

Interaction	Manufacturing Change: Decline		Growth		Total
A_1C_1	308	(51.08)	295	(48.92)	603
A_1C_2	91	(53.85)	78	(46.15)	169
A_1C_3	42	(42.86)	56	(47.14)	98
A_1C_4	25	(32.47)	52	(67.53)	77
A_2C_1	5	(27.78)	13	(72.22)	18
A_2C_2	191	(39.14)	297	(60.86)	488
A_2C_3	102	(37.92)	167	(62.08)	269
A_2C_4	72	(33.03)	146	(66.97)	218
A_3C_1	11	(30.56)	25	(69.44)	36
A_3C_2	239	(38.18)	387	(61.82)	626
A_3C_3	114	(40.71)	166	(59.29)	280
A_3C_4	74	(33.64)	146	(66.36)	220
Total	1274	(41.07)	1828	(58.93)	3102

	Expected Frequencies		Deviations From Observed Frequencies		Chi-Square Components	
	Decline	Growth	Decline	Growth	Decline	Growth
A_1C_1	248	355	+60	-60	14.52	10.14
A_1C_2	69	100	+22	-22	7.01	4.84
A_1C_3	40	58	+ 2	- 2	.10	.07
A_1C_4	32	45	- 7	+ 7	1.53	1.09
A_2C_1	7	11	- 2	+ 2	.57	.36
A_2C_2	200	288	- 9	+ 9	.40	.28
A_2C_3	110	159	- 8	+ 8	.58	.40
A_2C_4	90	128	-18	+18	3.60	2.53
A_3C_1	15	21	- 4	+ 4	1.07	.76
A_3C_2	257	369	-18	+18	1.26	.88
A_3C_3	115	165	- 1	+ 1	.01	.01
A_3C_4	91	129	-17	+17	3.18	2.24

X^2 (11) = 57.43 Sig. $<$ .001

TABLE 4.15
(cont.)

Interaction of Change in F. E. A. City Size (1950 - 1960) with Relative Location in Urban Field (1960) and Change in Number of Manufacturing Employees by Place of Work 1952 - 1962.

Interaction	Manufacturing Change: Decline		Manufacturing Change: Growth		Total
$a_1+a_2C_1+C_2$	425	(46.25)	494	(53.75)	919
$a_1+a_2C_3$	62	(41.61)	87	(58.39)	149
$a_1+a_2C_4$	61	(46.92)	69	(53.08)	130
$a_3C_1+C_2$	169	(39.58)	258	(60.42)	427
a_3C_3	101	(40.56)	148	(59.44)	249
a_3C_4	72	(37.11)	122	(62.89)	194
$a_4C_1+C_2$	250	(42.09)	344	(57.91)	594
a_4C_3	96	(38.55)	153	(61.45)	249
a_4C_4	38	(19.90)	153	(80.10)	191
Total	1274	(41.07)	1828	(58.93)	3102

	Expected Frequencies: Decline	Expected Frequencies: Growth	Deviations From Observed Frequencies: Decline	Deviations From Observed Frequencies: Growth	Chi-Square Components: Decline	Chi-Square Components: Growth
$a_1+a_2C_1+C_2$	377	542	+48	-48	6.11	4.25
$a_1+a_2C_3$	61	88	+ 1	- 1	.02	.01
$a_1+a_2C_4$	53	77	+ 8	- 8	1.21	.83
$a_3C_1+C_2$	175	252	- 6	+ 6	.21	.14
a_3C_3	102	147	- 1	+ 1	.01	.01
a_3C_4	80	114	- 8	+ 8	.80	.56
$a_4C_1+C_2$	244	350	+ 6	- 6	.15	.10
a_4C_3	102	147	- 6	+ 6	.35	.24
a_4C_4	80	111	-42	+42	22.05	15.89

X^2 (8)= 52.94 Sig.< .001

TABLE 4.16

Relative Accessibility (1950) Interaction with Relative Location in Urban Field (1960) and Change in Number of Manufacturing Employees by Place of Work 1952-1962.

Interaction	Manufacturing Change: Decline		Growth		Total
B_1C_1	204	(61.19)	130	(38.81)	335
B_1C_2	131	(52.40)	119	(47.60)	250
B_1C_3	32	(38.10)	52	(61.90)	84
B_1C_4	17	(17.52)	80	(82.48)	97
B_2C_1	73	(40.78)	106	(59.22)	179
B_2C_2	192	(39.92)	289	(60.08)	481
B_2C_3	108	(48.65)	114	(51.35)	222
B_2C_4	42	(29.37)	101	(70.63)	143
B_3C_1	46	(32.17)	97	(67.83)	143
B_3C_2	198	(35.87)	354	(64.13)	552
B_3C_3	118	(34.60)	223	(65.40)	341
B_3C_4	112	(40.73)	163	(59.27)	275)
Total	1274	(41.07)	1828	(58.93)	3102

	Expected Frequencies		Deviations From Observed Frequencies		Chi-Square Components	
	Decline	Growth	Decline	Growth	Decline	Growth
B_1C_1	137	198	+68	-68	33.75	23.35
B_1C_2	103	147	+28	-28	7.61	5.33
B_1C_3	34	60	- 2	+ 2	.12	.08
B_1C_4	40	57	-23	+23	13.22	9.28
B_2C_1	73	106	-	-	-	-
B_2C_2	198	283	- 6	+ 6	.18	.13
B_2C_3	91	131	+17	-17	3.18	2.21
B_2C_4	59	84	-17	+17	4.90	3.44
B_3C_1	59	84	-13	+13	2.86	2.01
B_3C_2	227	325	-29	+29	3.70	2.59
B_3C_3	140	201	-22	+22	3.46	2.41
B_3C_4	113	162	- 1	+ 1	.01	.01

X^2 (11)= 123.83 Sig.$<.001$

TABLE 4.16

Relative Accessibility Interaction with Relative Location in Urban Field (1960) and Change in Number of Manufacturing Employees by Place of Work 1952 - 1962.

Interaction	Decline		Growth		Total
B_1C_1	135	(57.69)	99	(42.31)	234
B_1C_2	68	(48.23)	73	(51.77)	141
B_1C_3	16	(34.04)	31	(65.96)	47
B_1C_4	15	(28.85)	37	(71.15)	52
B_2C_1	114	(57.58)	84	(42.42)	198
B_2C_2	183	(45.41)	220	(54.59)	403
B_2C_3	76	(51.35)	72	(48.65)	148
B_2C_4	27	(22.50)	93	(77.50)	120
B_3C_1	75	(33.33)	150	(67.67)	225
B_3C_2	270	(36.54)	469	(63.46)	739
B_3C_3	166	(36.73)	286	(63.27)	452
B_3C_4	129	(37.61)	214	(62.39)	343
Total	1274	(41.07)	1828	(58.93)	3102

	Expected Frequencies		Deviations From Observed Frequencies		Chi-Square Components	
	Decline	Growth	Decline	Growth	Decline	Growth
B_1C_1	96	138	+39	-39	15.84	11.02
B_1C_2	58	83	+10	-10	1.72	1.20
B_1C_3	19	28	- 3	+ 3	.47	.32
B_1C_4	21	31	- 6	+ 6	1.71	1.16
B_2C_1	81	117	+33	-33	13.44	9.31
B_2C_2	166	237	+17	-17	1.74	1.22
B_2C_3	61	87	+15	-15	3.69	2.59
B_2C_4	49	71	-22	+22	9.88	6.82
B_3C_1	92	133	-17	+17	3.14	2.17
B_3C_2	303	436	-33	+33	3.59	2.50
B_3C_3	186	266	-20	+20	2.15	1.50
B_3C_4	142	201	-13	+13	1.19	.84

X^2 (11)= 99.21 Sig.$<$ .001

TABLE 4.16

Interaction of Change in Relative Accessibility (1950 - 1960) with Relative Location in Urban Field (1960) and Change in Number of Manufacturing Employees by Place of Work 1952 - 1962.

Manufacturing Change

Interaction	Decline		Growth		Total
b_1C_1	261	(48.78)	274	(51.22)	535
b_1C_2	492	(41.07)	706	(58.93)	1198
b_2C_1	248	(41.13)	355	(58.97)	603
b_2C_2	166	(38.52)	265	(61.48)	431
b_3C_1	62	(50.82)	60	(49.18)	122
b_3C_2	29	(34.12)	56	(65.88)	85
b_4C_1	11	(25.00)	33	(75.00)	44
b_4C_2	5	(5.95)	79	(94.05)	84
Total	1274	(41.07)	1828	(58.93)	3102

	Expected Frequencies		Deviations From Observed Frequencies		Chi-Square Components	
	Decline	Growth	Decline	Growth	Decline	Growth
b_1C_1	202	315	+41	-41	7.64	5.34
b_1C_2	492	706	--	--	--	--
b_2C_1	248	355	--	--	--	--
b_2C_2	177	254	-11	+11	.68	.48
b_3C_1	50	72	+12	-12	2.88	2.00
b_3C_2	35	50	- 6	+ 6	1.03	.72
b_4C_1	18	26	- 7	+ 7	2.72	1.88
b_4C_2	34	50	-29	+29	24.73	16.82

X

acting with B_1 and B_2 is evident, too, producing 24.9 per cent of the overall chi-square value. The most interesting phenomenon is the low interaction effect in the B_3C groups, particularly in that group that was the most accessible and intensely urbanized in 1950 - B_3C_4.

With the improvement in relative accessibility by 1960, the relationships described above change. The negative effect is not concentrated in the B_1C_1 and B_1C_2 groups but spreads to all B_2C groups except B_2C_4, and the positive effect of the interaction B_1C_4 although it still exists, is weakened. Generally, by 1960 the importance of the B_iC_i interaction has declined as compared to the 1950 levels. Once again, it appears that the interaction between the relative location components is not very powerful as a source of deviations from the national proportions of counties experiencing growth and decline. In turn, this conclusion implies that, with respect to the change in manufacturing employment by place of work, the processes of hierarchical diffusion, dispersion and decentralization are more "complete" than with respect to change in number of manufacturing establishments. Such a conclusion could explain why the dynamic elements (the change in FEA city population, the change in population potential, their interaction, and the interaction of each of them with location in the urban field) are relatively more important with respect to this employment variable than they were with respect to the change in the number of establishments. Also, the mobility of labor and the immobility of plants once they are established could contribute to the results.

Relative Location and Change Differentials

The relative insignificance of initial relative location components for the rate of change in employment in manufacturing by place of work is revealed in the analysis of variance summarized in Table

4.17. It is the interaction effects BC and AC that are probably the most significant in affecting deviations from the average change in manufacturing employment by place of work. The least squares estimates of the directions of the interaction effects reveal that the BC_4 interaction is positive except for B_3C_4, which by induction is negative. The effect of the B_3C interaction, except B_3C_4, also is positive. All other interactions are negative except that for B_1C_2. The direction of the interaction effect BC on actual change in employment is thus only partly consistent with the deviations of the chi-square analysis. Particularly, note the negative effect of B_3C_4 group in 1950 did not deviate substantially from the national proportion but there their rates were probably lower than those in the B_1C_4 and B_2C_4 groups. Alternatively the decline was greater in counties belonging to the B_3C_4 group than the B_1C_4 and B_2C_4 groups. A combination of excess decline and slower growth could be responsible for the negative effect. In any event it becomes evident that having high relative accessibility in 1950 and being close to an FEA city was not an "asset" in terms of growth of manufacturing employment. It probably indicates a relative decline in manufacturing employment by place of work for the most accessible counties of the country in 1950. In fact, such a result confirms the suspicion that dispersion and decentralization affected the spread of change in manufacturing employment more than they did the change in the number of manufacturing establishments. On the other hand, the AC_4 deviations reveal that A_3C_4 had a highly positive effect on change in employment by place of work. The large cities of 1950 and their suburban zones are characterized by rapid increase in manufacturing employment by place of work.

Seemingly, the positive A_iC_i effect and the negative B_iC_i effect contradict each other. It could therefore be that the negative effect of B_iC_i is concentrated in urban fields of small and medium central cities in 1950.

TABLE 4.17

Analysis of Variance 1950

Effect	Mean Square	Degrees of Freedom	F. Ratio	Significance
A	15.18	2 ,3066	2.41	<.0894
B	14.08	2 ,3066	2.24	<.1067
C	16.62	3 ,3066	2.63	<.0475
AB	13.66	4 ,3066	2.17	<.0697
AC	9.54	6 ,3066	1.51	<.1678
BC	21.34	6 ,3066	3.39	<.0025
ABC	6.07	12 ,3066	.96	<.4822

ANOVA: Each Effect Alone.

Effect	Mean Square	Degrees of Freedom	F. Ratio	Significance
A	15.18	2 ,3066	2.41	<.0894
B	23.65	2 ,3066	3.76	<.0234
C	25.42	3 ,3066	4.04	<.0071
AB	14.81	4 ,3066	2.35	<.0517
AC	22.24	6 ,3066	3.53	<.0018
BC	30.46	6 ,3066	4.84	<.0001
ABC	10.59	12 ,3066	1.68	<.0642

ANOVA: Significant Effects Only

Effect	Mean Square	Degrees of Freedom	F. Ratio	Significance
BC	30.46	6 ,3066	4.84	<.0001
AC	13.28	6 ,3066	2.11	<.0487

Least Square Estimates of Effects

General Mean	+1.67
B_1C_1	- .89
B_1C_2	- .32
B_1C_3	-1.04
B_2C_1	- .75
B_2C_2	- .64
B_2C_3	- .63
A_1C_1	- .32
A_1C_2	- .45
A_1C_3	+ .16

The results of the analysis of variance from the 1960 relative location perspective are similar to those of the 1950 analysis (Table 4.18). Once again, the main effects are not as significant. Only the two interaction effects AC and AB are statistically significant when all other factors are ignored and remain so when other factors are eliminated. The direction of the effects is given by the coefficients of the least-squares estimates. The BC effect is quite familiar: B_1 and B_2 exert a negative effect in all commuting zones except C_4, the suburban or intense commuting areas of urban fields, and B_3 exerts positive effect in all commuting zones. Thus, the deviation noted in the 1950 analysis had been "corrected" by 1960 with the general shift and improvement in relative accessibility. It is noticeable that the AC effects are more consistent between 1950 and 1960. On the other hand, there are no systematic relationships within the interaction groups, and no possibility of a generalization about the interaction effect. The predictions based on either 1960 relative location or on 1950 relative location are quite accurate, however.

All dimensions of change in relative location between 1950 and 1960 except the ab interaction effect are statistically significant when they appear in analysis of variance alone (see Table 4.19). When all significant effects in Table 4.19 are introduced into the analysis in order of their significance, the end result is that the main effects (a, change in FEA population 1950-1960; and b, change in county population potential 1950-1960) and the interaction effect bC (the change in relative accessibility together with the location in the urban field in 1960) are statistically significant.

The influence is in the direction expected from the parallel chi-square analysis: a_1-strong negative effect; a_3-population growth less than 20 per cent, a weak negative effect; a_4-rapid population growth in the FEA city, positive effect. Rapid change in

TABLE 4.18

Analysis of Variance (1960)

Effect	Mean Square	Degrees of Freedom	F. Ratio	Significance
A	28.55	2 ,3066	4.51	<.0110
B	14.85	2 ,3066	2.35	<.0951
C	9.21	3 ,3066	1.46	<.2229
AB	13.73	4 ,3066	2.17	<.0693
AC	9.67	6 ,3066	1.53	<.6330
BC	16.12	6 ,3066	2.55	<.0134
ABC	1.07	12 ,3066	.17	<.8863
ANOVA: Each Effect Alone				
A	28.55	2 ,3066	4.51	<.0110
B	25.40	2 ,3066	4.02	<.0180
C	25.42	3 ,3066	4.02	<.0072
AB	8.62	4 ,3066	1.36	<.2426
AC	20.84	6 ,3066	3.29	<.0031
BC	18.07	6 ,3066	2.86	<.0089
ABC	5.04	12 ,3066	.80	<.6539
ANOVA: Significant Effects Only				
AC	20.84	6 ,3066	3.29	<.0031
BC	17.02	6 ,3066	2.69	<.0131

Least Square Estimates of Effects

General Mean	+1.66
A_1C_1	- .55
A_1C_2	- .43
A_1C_3	- .12
A_2C_1	+ .33
A_2C_2	+ .64
A_2C_3	+ .50
B_1C_1	- .75
B_1C_2	-1.04
B_1C_3	- .12
B_2C_1	- .99
B_2C_2	- .30
B_2C_3	- .82

TABLE 4.19

Analysis of Variance (1950-1960)

Effect	Mean Square	Degrees of Freedom	F. Ratio	Significance
a	32.64	3 ,3074	5.17	<.0015
b	52.08	1 ,3074	8.24	<.0041
C	3.90	3 ,3074	.62	<.6057
ab	18.35	3 ,3074	2.90	<.0332
aC	4.08	9 ,3074	.65	<.7592
bC	10.97	3 ,3074	1.74	<.1561
abC	N.A.	9 ,3074	N.A.	N.A.
ANOVA: Each Effect Alone				
a	32.64	3 ,3074	5.17	<.0015
b	37.72	1 ,3074	5.97	<.0146
C	25.42	3 ,3074	4.02	<.0072
ab	6.95	3 ,3074	1.10	<.3447
aC	17.35	9 ,3074	2.75	<.0411
bC	12.49	3 ,3074	1.98	<.0378
abC	14.77	9 ,3074	2.34	<.0126
ANOVA: Significant Effects Only				
a	32.64	3 ,3074	5.17	<.0015
b	52.08	1 ,3074	8.24	<.0041
bC	23.29	3 ,3074	3.69	<.0114

Least Square Estimates of Effects

General Mean	+1.68
a_1	- .68
a_2	- .02
a_3	- .04
b_1	- .51
b_1C_1	+1.10
b_1C_2	+ .64
b_1C_3	+ .60

population potential had a positive effect. Further, b_1c exerted a positive effect, except for b_1c_4.

One can conclude that relative accessibility change, population change in central cities of economic areas, and the interaction between accessibility change and location in the urban field were all clearly associated with the change in manufacturing employment by place of work.

Finally, the analysis of variance presented Table 4.20 shows that what influenced the change in manufacturing employment by place of work during the period was not only the static or dynamic relative location as was revealed earlier, but a combination of the initial relative location and the change in its components. In particular, it was the initial relative accessibility and the change in relative accessibility that were the most significant. However, the initial size of the FEA city and its change also were significantly associated with manufacturing change. <u>These two interactions represent the most powerful spatial changes in the United States during the 1950's</u>. After the Aa and Bb effects are eliminated from the analysis, location within urban fields loses significance, perhaps because it had to be treated in the analysis as a spatial constant unaccompanied by a dynamic element. The results demonstrate that change in manufacturing employment by place of work was affected by the hierarchical level of the FEA and change in the population of the FEA city, clearly an hierarchical effect. Even more obvious is the dependency of employment change on rapid change in relative accessibility, a clear indication of a dispersion effect.

TABLE 4.20

Analysis of Variance 1950 and 1950-1960

Effect	Mean Square	Degrees of Freedom	F. Ratio	Significance
Aa	18.87	6 ,2959	3.00	<.0063
Bb	32.47	2 ,2959	5.17	<.0058
C	3.06	3 ,2959	.49	<.6918
AaBb	4.56	12 ,2959	.73	<.7268
AaC	4.76	18 ,2959	.76	<.7532
BbC	5.70	6 ,2959	.91	<.4884
ANOVA: Each Effect Alone				
Aa	18.87	6 ,2959	3.00	<.0063
Bb	45.85	2 ,2959	7.30	<.0007
C	25.42	3 ,2959	4.05	<.0070
AaBb	4.79	12 ,2959	.76	<.6900
AaC	9.20	18 ,2959	1.46	<.0924
BbC	12.16	6 ,2959	1.94	<.0714
ANOVA: Significant Effects Only				
Bb	45.85	2 ,2959	7.30	<.0007
Aa	14.40	6 ,2959	2.29	<.0327

Least Square Estimates of Effects

General Mean	+1.57
B_1b_1	- .77
B_2b_1	- .48
A_1a_1	- .49
A_1a_2	+ .28
A_1a_3	+ .14
A_2a_1	+ .44
A_2a_2	+ .11
A_2a_3	+ .29

CHAPTER 5

CHANGES IN THE LOCATION OF LARGE- AND MEDIUM-SCALE INDUSTRY

LARGE-SCALE INDUSTRIAL ESTABLISHMENTS

The locational requirements of large industrial establishments, those that employ more than 99 workers, vary significantly from establishments with smaller numbers of employees. First, large establishments usually require a larger market than smaller ones. While smaller plants are primarily associated with local markets and only to a limited degree with regional and national markets, the principal market of large plants is likely to be regional or national in character rather than local. Second, large establishments require a larger land area for their current operation and foreseeable future expansion. Such a requirement implies that new large plants are likely to locate further away from central cities, provided that the required labor force is available. Thirdly, the infra-structure requirements of large establishments as against those of small or medium size plants result either in greater attachment to economic areas with strong urban centers, or in the realization of significant internal economies that negates the need for spatial centralization. Finally, given the technological changes in industry that have resulted in reductions in labor-force requirements, and the large investment needs of the modern capital-intensive establishments, the growth in the number of such plants is likely to be very selective geographically. These considerations suggest that more than one spatial process may be associated with the change in such establishments.

Hierarchical Diffusion

The relative location of counties in the urban hierarchy in both 1950 and in 1960 is associated with the change in total number of large manufacturing establishments (Table 5.1). The growth proportions remained essentially constant in these years, varying sharply with FEA size, although counties' growth prospects did change as a result of relative location changes. The fact that only 685 counties experienced any growth at all in large-scale manufacturing plants (282 of these in the largest FEA's in 1950, rising to 321 by 1960) shows how selective this form of manufacturing development is. The fact that 146 counties, as measured by 1950 status, and 114 by 1960 status, had large-scale manufacturing growth--or that 235 counties located within stagnant or declining FEA's also had growth of large-scale manufacturing establishments--indicates the complexity of the processes involved, and that diffusion was taking place, although the rate of diffusion was not materially shifted in the decade.

Dispersion

Similar conclusions arise in the case of dispersion. Growth of large-scale manufacturing varied steeply with respect to relative accessibility in both 1950 and 1960, although heartland expansion di improve the growth prospects of many counties during the decade (Tab 5.2). About three out of five counties that experienced growth between 1952 and 1962 possessed high population potential in 1950, and of those counties experiencing growth in the decade under study, 74.45 per cent had achieved high levels of relative accessibility by 1960. These facts imply several things. In order to experience growth in numbers of large manufacturing establishments, counties ha to possess high relative accessibility to begin with, or possess the

TABLE 5.1

Urban Hierarchy (1950) and Change in the Total Number of Manufacturing Establishments with more than 99 Employees 1952 - 1962.

Manufacturing Change	A_1		A_2		A_3		Total	P
Decline	922	(86.33)	762	(74.78)	733	(72.22)	2417	(77.92)
Growth	146	(13.67)	257	(25.22)	282	(27.78)	685	(22.08)
Total	1068		1019		1015		3102	

	Expected Frequencies			Deviations From Observed Frequencies			Chi-Square Components		
	A_1	A_2	A_3	A_1	A_2	A_3	A_1	A_2	A_3
Decline	832	794	791	+90	-32	-58	7.74	1.29	4.25
Growth	236	225	224	-90	+32	+58	34.32	4.55	15.02

X^2 (2)= 67.17 Sig. < .001

Urban Hierarchy (1960) and Change in the Total Number of Manufacturing Establishments with more than 99 Employees 1952 - 1962.

Manufacturing Change	A_1		A_2		A_3		Total	P
Decline	833	(87.96)	743	(74.82)	841	(72.38)	2417	(77.92)
Growth	114	(12.04)	250	(25.18)	321	(27.62)	685	(22.08)
Total	947		993		1162		3102	

	Expected Frequencies			Deviations From Observed Frequencies			Chi-Square Components		
	A_1	A_2	A_3	A_1	A_2	A_3	A_1	A_2	A_3
Decline	738	774	905	+95	-31	-64	12.23	1.24	4.52
Growth	209	219	257	-95	+31	+64	43.18	4.39	15.94

X^2 (2)= 81.50 Sig. < .001

Population Change in the Urban Hierarchy 1950 - 1960 and Change in the Total Number of Manufacturing Establishments With More Than 99 Employees 1952 - 1962.

Manufacturing Change	$a_1 + a_2$		a_3		a_4		Total	
Decline	963	(80.38)	682	(78.39)	772	(74.66)	2417	(77.92)
Growth	235	(19.62)	188	(21.61)	262	(25.34)	685	(22.08)
Total	1198		870		1034		3102	

	Expected Frequencies			Deviations From Observed Frequencies			Chi-Square Components		
	a_1+a_2	a_3	a_4	a_1+a_2	a_3	a_4	a_1+a_2	a_3	a_4
Decline	933	678	806	+30	+4	-34	.96	.02	1.43
Growth	265	192	228	-30	-4	+34	3.40	.08	5.08

X^2 (2)= 10.97 Sig. < .010

TABLE 5.2

Relative Accessibility (1950) and Change in the Total Number of Manufacturing Establishments with More Than 99 Employees 1952 - 1962.

Manufacturing Change	B_1		B_2		B_3		Total	P
Decline	675	(88.12)	829	(80.88)	913	(69.64)	2417	(77.92)
Growth	91	(11.88)	196	(19.12)	398	(30.36)	685	(22.08)
Total	766		1025		1311		3102	

	Expected Frequencies			Deviations From Observed Frequencies			Chi-Square Components		
	B_1	B_2	B_3	B_1	B_2	B_3	B_1	B_2	B_3
Decline	597	799	1021	+78	+30	-108	10.19	1.13	11.42
Growth	169	226	290	-78	-30	+108	36.00	3.98	40.22

X^2 (2)= 102.94 Sig. < .001

Relative Accessibility (1960) and Change in Total Number of Manufacturing Establishments with More than 99 Employees 1952 - 1962.

Manufacturing Change	B_1		B_2		B_3		Total
Decline	417	(87.97)	751	(86.42)	1249	(71.01)	2417
Growth	57	(12.03)	118	(13.58)	510	(28.99)	685
Total	474		869		1759		3102

	Expected Frequencies			Deviations From Observed Frequencies			Chi-Square Components		
	B_1	B_2	B_3	B_1	B_2	B_3	B_1	B_2	B_3
Decline	369	677	1371	+48	+74	-122	6.24	8.09	10.86
Growth	105	192	388	-48	-74	+122	21.94	28.52	38.36

X^2 (2)= 114.01 Sig. < .001

Change in Relative Accessibility (1950 - 1960) and Change in Total Number of Manufacturing Establishments With More Than 99 Employees 1952 - 1962

Manufacturing Change	b_1		b_2		Total	P
Decline	2165	(78.24)	252	(75.22)	2417	(77.92)
Growth	602	(21.76)	83	(24.78)	685	(22.08)
Total	2767		335		3102	

	Expected Frequencies		Deviations From Observed Frequencies		Chi-Square Components	
	b_1	b_2	b_1	b_2	b_1	b_2
Decline	2156	261	+9	-9	.04	.31
Growth	611	74	-9	+9	.13	1.09

X^2 (1)= 1.57 Sig. < .300

potential to achieve such a high level within a short period of time. Dispersion of industrial activity to counties with a low level of relative accessibility in 1950 could be seen as a result of the expected increase in relative accessibility of such counties. The relatively small numbers of counties that grew and remained in the low- and medium-potential categories (17.23 and 8.32 per cent, respectively) could indicate those places where expected relative accessibility change did not materialize, or materialized only to a limited degree. Such counties could include those where interstate highway programs were not completed on time, or counties where the predicted growth in the size of the market did not occur. But on the other hand, such counties could be those where labor force availability or raw materials were the crucial factors in the location decision. But these factors aside, the high proportion of counties that experienced growth in numbers of large manufacturing establishments, and that possessed high relative accessibility in 1960, indicates once again, that initial relative location together with the change in such accessibility is the major factor in explaining the dispersion of industrial activity.

<u>Decentralization with Urban Fields</u>

So far, the chi-square analysis has revealed only limited hierarchical diffusion and dispersion of large manufacturing establishments; of more importance is hierarchical growth and heartland expansion. Table 5.3 reveals the extent of decentralization. Growth is found in all zones of commuting around central cities. Of the 685 counties experiencing growth 98 were located beyond the outer edge of daily commuting within urban fields in 1960, 234 in the next commuting zone, 157 in the commuting zone with between 5 and 20 per cent of the resident labor force to the central city and the rest, 196 in the innermost aone of suburbanization. Growth did extend

TABLE 5.3

Relative Location in Urban Fields (1960) and Change in Total Number of Manufacturing Establishments with More Than 99 Employees 1952 - 1962.

Manufacturing Change

	C_1		C_2		C_3		C_4		Total	P
Decline	559	(85.08)	1049	(81.76)	490	(75.73)	319	(61.94)	2417	(77.92)
Growth	98	(14.92)	234	(18.24)	157	(24.27)	196	(38.06)	685	(22.08)
Total	657		1283		647		515		3102	

	Expected Frequencies				Deviations From Observed Frequencies				Chi-Square Components			
	C_1	C_2	C_3	C_4	C_1	C_2	C_3	C_4	C_1	C_2	C_3	C_4
Decline	512	1000	504	401	+47	+49	-14	-82	4.31	2.40	.39	16.77
Growth	145	283	143	114	-47	-49	+14	+82	15.23	8.48	1.37	58.98

X^2 (3)= 107.93 Sig.< .001

outwards to the furthest peripheries, then, but the growth proportion dropped significantly, from 38.06 in the core to 14.92 per cent in the periphery, showing that the decentralization trend towards the periphery of urban fields during the fifties, although clearly evident, was highly selective. The other side of the picture is that, the closer a county is to the urban center of an economic area, the greater are its chances of experiencing growth in number of large-scale manufacturing establishments.

Hierarchical Diffusion and Dispersion

The interaction of size of urban centers with population potentials reveals that, for any hierarchical level, the higher the relative accessibility of counties, the greater the proportion of counties experiencing growth in total number of large manufacturing plants. The growth proportions vary from 6.92 per cent in group A_1B_1 to 27.88 per cent in group A_1B_3, from 20.12 per cent in A_2B_1 to 32.01 in A_3B_3. Groups A_2B_3 and A_3B_3 contained almost one out of two counties that experienced growth. These proportions refer to the relative location of counties in the initial state (1950, Table 5.4). When the distribution of growth is examined for the 1960 conditions, both the generalization and the particulars still hold. However, the concentration of the growth in counties belonging to A_2B_3 and A_3B_2 is much stronger: 63 per cent of all counties experiencing growth belonged to these groups. Almost two out of five growth counties were located in 1960 within the A_3B_3 group. Yet in the most accessible areas of the country associated with the lowest level of the urban hierarchy there were still to be found 11 per cent of the growing counties. Impulses of growth, although much weaker, did filter to all other groups in all locations. But the fact that growth is mostly associated with the achievement of high relative accessibility by 1960 leads to the conclusion that the crucial factor affecting the location of new large

TABLE 5.4

Urban Hierarchy Interaction with Relative Accessibility (1960) and Change in Total Number of Manufacturing Establishments with more than 99 employees 1952 - 1962.

Interaction	Manufacturing Change: Decline		Growth		Total
A_1B_1	270	(92.78)	21	(7.22)	291
A_1B_2	320	(94.67)	18	(5.33)	338
A_1B_3	243	(76.42)	75	(23.58)	318
A_2B_1	89	(79.46)	23	(20.54)	112
A_2B_2	197	(81.74)	44	(18.26)	241
A_2B_3	457	(71.41)	183	(28.59)	640
A_3B_1	58	(81.69)	13	(18.31)	71
A_3B_2	234	(80.69)	56	(19.31)	290
A_3B_3	549	(68.54)	252	(31.46)	801
Total	2417	(77.92)	685	(20.08)	3102

	Expected Frequencies		Deviations From Observed Frequencies		Chi-Square Components	
	Decline	Growth	Decline	Growth	Decline	Growth
A_1B_1	227	64	+43	-43	8.14	28.99
A_1B_2	263	75	+57	-57	12.35	43.32
A_1B_3	248	70	- 5	+ 5	.10	.36
A_2B_1	87	25	+ 2	- 2	.05	.16
A_2B_2	188	53	+ 9	- 9	.43	1.53
A_2B_3	499	141	-42	+42	3.54	12.51
A_3B_1	55	16	+ 3	- 3	.16	.56
A_3B_2	226	64	+ 8	- 8	.28	1.00
A_3B_3	624	177	-75	+75	9.01	31.78

X^2 (8)= 154.17 Sig. .001

Interaction of Change in F.E.A. City Size with Change in Relative Accessibility (1950 - 1960) and Change in Total Numbers of Manufacturing Establishments with more than 99 Employees 1952 - 1962.

Interaction	Manufacturing Change: Decline		Growth		Total
a_1b_1	424	(86.35)	77	(13.65)	491
a_1b_2	110	(95.65)	5	(4.35)	115
a_2b_1	414	(72.76)	155	(27.24)	569
a_2b_2	15	(65.22)	8	(34.78)	23
a_3b_1	644	(79.02)	171	(20.98)	815
a_3b_2	38	(69.09)	17	(30.91)	55
a_4b_1	683	(76.57)	209	(23.43)	892
a_4b_2	89	(62.68)	53	(37.32)	142
Total	2417	(77.92)	685	(20.08)	3102

	Expected Frequencies		Deviations From Observed Frequencies		Chi-Square Components	
	Decline	Growth	Decline	Growth	Decline	Growth
a_1b_1	383	108	+41	-41	4.39	15.56
a_1b_2	90	25	+20	-20	4.44	16.00
a_2b_1	443	126	-29	+29	1.90	6.67
a_2b_2	18	5	- 3	+ 3	.50	1.80
a_3b_1	635	180	+ 9	- 9	.13	.45
a_3b_2	43	12	- 5	+ 5	.58	2.08
a_4b_1	695	197	-12	+12	.21	.73
a_4b_2	110	32	-21	+21	4.01	13.78

X^2 (7)= 73.23 Sig.< .001

TABLE 5.4
(cont.)

Urban Hierarchy Interaction with Relative Accessibility (1950) and the Change in Total Number of Manufacturing Establishments with more than 99 employees 1952 - 1962.

Interaction	Decline		Growth		Total
A_1B_1	457	(33.08)	34	(6.92)	491
A_1B_2	302	(86.04)	49	(13.96)	351
A_1B_3	163	(72.12)	63	(27.88)	226
A_2B_1	135	(79.88)	34	(20.12)	169
A_2B_2	287	(77.99)	81	(22.01)	368
A_2B_3	340	(70.54)	142	(29.46)	482
A_3B_1	83	(78.30)	23	(21.70)	106
A_3B_2	240	(78.43)	66	(21.57)	306
A_3B_3	410	(67.99)	193	(32.01)	603
Total	2417	(77.92)	685	(22.08)	3102

	Expected Frequencies		Deviations From Observed Frequencies		Chi-Square Components	
	Decline	Growth	Decline	Growth	Decline	Growth
A_1B_1	383	108	+74	-74	14.30	50.70
A_1B_2	273	78	+29	-29	3.08	10.78
A_1B_3	176	50	-13	+13	.96	3.38
A_2B_1	132	37	+ 3	- 3	.07	.24
A_2B_2	287	81	---	---	-----	-----
A_2B_3	376	106	-36	+36	3.45	12.23
A_3B_1	83	23	---	---	-----	-----
A_3B_2	238	68	+ 2	- 2	.02	.06
A_3B_3	469	134	-59	+59	7.42	25.98

X^2 (8)= 132.67 Sig. .001

manufacturing plants is indeed high relative accessibility more than it is the association with high hierarchical levels.

Hierarchical Diffusion and Decentralization

Growth in the number of large establishments occurred in all commuting zones at all hierarchical levels (Table 5.5). The greatest growth rates were found in two zones, however: the immediate suburbanization zones (A_2C_4 and A_3C_4) and in the furthest peripheral zones (A_2C_1 and A_3C_1). With respect to the lowest level of the urban hierarchy (A_1), the highest growth proportion tended to be close to the FEA central city in 1950, but this effect vanished by 1960. In counties associated with the high-level FEA's, however, the growth rates peaked in both the most central and the most peripheral counties. During the decade, large plants thus were being built either in central or peripheral locations throughout the country. To the extent that there was a particular pattern to the growth, about 39 per cent of the growing counties were in three zones in 1950 (A_3C_2, A_3C_3 and A_3C_4). This figure jumped to about 46 per cent in 1960. This concentration indicates a growing tendency for large manufacturing plants to locate in the largest economic areas, but close to the outer margin of daily commuting.

The interaction between FEA city population change and the relative location of counties within their urban field shows that 2 out of five counties experiencing growth were associated with very fast growth in these FEA city populations. At the other extreme, one out of five counties were located in an economic area where the FEA city either experienced no population change or declined. Most of the chi-square values are concentrated in these groups, and the deviation pattern shows a negative deviation for the $a_1+a_2C_1+C_2$ group and a positive one for A_4C_4. These two groups contribute fifty per cent the chi-square value. These observations, although independent of

TABLE 5.5

Urban Hierarchy (1950) Interaction with Relative Location in Urban Field (1960) and Change in Total Number of Manufacturing Establishments With More than 99 Employees 1952 - 1962.

Interaction	Manufacturing Change: Decline		Manufacturing Change: Growth		Total
A_1C_1	534	(88.56)	71	(11.44)	605
A_1C_2	209	(92.48)	17	(7.52)	226
A_1C_3	102	(79.07)	27	(20.93)	129
A_1C_4	77	(71.30)	31	(28.70)	108
A_2C_1	7	(41.18)	10	(58.82)	17
A_2C_2	419	(82.48)	89	(17.52)	508
A_2C_3	206	(74.64)	70	(25.34)	276
A_2C_4	130	(59.63)	88	(40.37)	218
A_3C_1	18	(51.43)	17	(48.57)	35
A_3C_2	421	(76.68)	128	(23.32)	549
A_3C_3	182	(75.21)	60	(24.79)	242
A_3C_4	112	(59.26)	77	(40.74)	189
Total	2417	(77.92)	685	(22.08)	3102

	Expected Frequencies		Deviations From Observed Frequencies		Chi-Square Components	
	Decline	Growth	Decline	Growth	Decline	Growth
A_1C_1	471	134	+63	-63	8.43	29.62
A_1C_2	176	50	+33	-33	6.19	21.78
A_1C_3	100	29	+ 2	- 2	.04	.14
A_1C_4	84	24	- 7	+ 7	.58	2.04
A_2C_1	13	4	- 6	+ 6	2.77	2.67
A_2C_2	396	112	+23	-23	1.34	4.72
A_2C_3	215	60	- 9	+ 9	..38	1.35
A_2C_4	170	48	-40	+40	9.41	33.33
A_3C_1	27	8	- 9	+ 9	3.00	10.12
A_3C_2	428	121	- 7	+ 7	.11	.40
A_3C_3	189	53	- 7	+ 7	.26	.92
A_3C_4	148	41	-36	+36	8.76	31.61

X^2 (11)= 179.97 Sig.<.001

TABLE 5.5
(cont.)

Urban Hierarchy Interaction with Relative Location in Urban Fields (1960) and Change in Total Number of Manufacturing Establishments With More Than 99 Employees 1952 - 1962.

Interaction	Manufacturing Change: Decline		Growth		Total
A_1C_1	533	(88.39)	70	(11.61)	603
A_1C_2	158	(93.49)	11	(6.51)	169
A_1C_3	79	(80.61)	19	(19.31)	98
A_1C_4	63	(81.82)	14	(18.18)	77
A_2C_1	7	(38.89)	11	(61.11)	18
A_2C_2	407	(83.40)	81	(16.60)	488
A_2C_3	204	(75.84)	65	(24.16)	269
A_2C_4	125	(57.34)	93	(42.64)	218
A_3C_1	19	(52.78)	17	(47.22)	36
A_3C_2	484	(77.32)	142	(22.68)	626
A_3C_3	207	(73.93)	73	(26.07)	280
A_3C_4	131	(59.55)	97	(40.45)	220
Total	2417	(77.92)	685	(22.08)	3102

	Expected Frequencies		Deviations From Observed Frequencies		Chi-Square Components	
	Decline	Growth	Decline	Growth	Decline	Growth
A_1C_1	470	133	+63	-63	8.44	29.84
A_1C_2	132	37	+26	-26	5.12	18.27
A_1C_3	76	22	+ 3	- 3	.12	.41
A_1C_4	60	17	+ 3	- 3	.15	.53
A_2C_1	14	4	- 7	+ 7	3.50	12.25
A_2C_2	380	108	+27	-27	1.92	6.75
A_2C_3	210	59	- 6	+ 6	.17	.61
A_2C_4	170	48	-45	+45	11.91	42.19
A_3C_1	28	8	- 9	+ 9	2.89	10.12
A_3C_2	488	138	- 4	+ 4	.03	.12
A_3C_3	218	62	-11	+11	.56	1.95
A_3C_4	171	49	-40	+40	9.36	32.65

X^2 (11)= 199.86 Sig.< .001

TABLE 5.5
(cont.)

Interaction of Change in F.E.A. City Size (1950 - 1960) with Relative Location in Urban Field (1960) and Change in Total Number of Manufacturing Establishments with More Than 99 Employees 1952 - 1962.

Interaction	Manufacturing Change: Decline		Growth		Total
$a_1+a_2C_1+C_2$	771	(83.90)	148	(16.10)	919
$a_1+a_2C_3$	110	(73.83)	39	(26.17)	149
$a_1+a_2C_4$	82	(63.08)	45	(36.92)	130
$a_3C_1+C_2$	346	(81.03)	81	(18.97)	427
a_3C_3	202	(81.12)	47	(18.88)	249
a_3C_4	134	(69.07)	60	(30.93)	194
$a_4C_1+C_2$	491	(82.66)	103	(17.34)	594
a_4C_3	178	(71.49)	71	(28.51)	249
a_4C_4	103	(53.93)	88	(46.07)	191
Total	2417	(77.92)	685	(22.08)	3102

	Expected Frequencies		Deviations From Observed Frequencies		Chi-Square Components	
	Decline	Growth	Decline	Growth	Decline	Growth
$a_1+a_2C_1+C_2$	716	203	+55	-55	4.22	14.90
$a_1+a_2C_3$	116	33	- 6	+ 6	.31	1.09
$a_1+a_2C_4$	101	29	-19	+19	3.57	12.45
$a_3C_1+C_2$	333	94	+13	-13	.51	1.80
a_3C_3	194	55	+ 8	- 8	.33	1.16
a_3C_4	151	43	-17	+17	1.91	6.72
$a_4C_1+C_2$	463	131	+28	-28	2.98	5.98
a_4C_3	194	55	-16	+16	1.32	4.65
a_4C_4	149	42	-46	+46	14.20	50.38

X^2 (8) = 128.48 Sig. < .001

city size, indicate that decentralization took place regardless of the degree of change in the FEA city population. On the other hand, positive deviations from the expected proportion are always associated with areas of intense commuting and negative deviations with marginal location in urban fields.

Dispersion and Decentralization

Earlier it was shown that counties experiencing growth in the number of large manufacturing establishments either had high relative accessibility in 1950 or have achieved it by 1960. Fifty-eight per cent of the growing counties were in the B_3 group in 1950 and about 75 per cent in 1960. These facts are important in the context of interaction between relative accessibility and location within the urban field, serving to emphasize that dispersion as a result of improved relative accessibility could be the crucial factor in attracting new large manufacturing plants. Given the importance or relative accessibility, location within urban fields does make a difference (Table 5.6). The chi-square analysis shows that of all counties having low or medium levels of population potential in 1950 or in 1960, only the groups of counties that by 1960 were located closest to the central FEA city deviated positively from expected proportions. In other words, groups B_1C_4 and B_2C_4 had larger than expected proportions of counties experiencing growth, while all other groups showed negative or insignificant deviations.

The four interaction groups B_3C had positive deviations. That is, high relative potential in 1950 or in 1960 was associated with accelerated growth in a number of large plants. High relative accessibility simultaneously implies not only dispersion, but also accelerated geographic decentralization. In fact, the average proportion of counties experiencing growth in the B_3C groups was 31.82 per cent in 1950 and 30.15 in 1960, substantially higher than the national

TABLE 5.6

Relative Accessibility (1950) Interaction with Relative Location In Urban Field (1960) and in Change in Total Number of Manufacturing Establishments with more than 99 Employees 1950 - 1962.

Interaction	Manufacturing Change: Decline		Growth		Total
B_1C_1	318	(94.92)	17	(5.08)	332
B_1C_2	236	(94.40)	14	(5.60)	250
B_1C_3	67	(79.76)	17	(20.24)	84
B_1C_4	54	(55.67)	43	(44.33)	97
B_2C_1	149	(83.24)	30	(16.76)	179
B_2C_2	410	(85.24)	71	(14.76)	481
B_2C_3	187	(84.23)	35	(15.77)	222
B_2C_4	83	(58.04)	60	(41.96)	143
B_3C_1	92	(64.34)	51	(35.66)	143
B_3C_2	403	(73.01)	149	(26.99)	552
B_3C_3	236	(69.21)	105	(30.79)	341
B_3C_4	182	(66.18)	93	(33.82)	275
Total	2417	(77.92	685	(22.08)	3102

	Expected Frequencies		Deviations From Observed Frequencies		Chi-Square Components	
	Decline	Growth	Decline	Growth	Decline	Growth
B_1C_1	261	74	+57	-57	12.45	43.90
B_1C_2	195	55	+41	-41	8.62	30.56
B_1C_3	65	19	+ 2	- 2	.06	.21
B_1C_4	76	21	-22	+22	6.37	23.05
B_2C_1	139	40	+10	-10	.72	2.50
B_2C_2	375	106	+35	-35	3.27	11.56
B_2C_3	173	49	+14	-14	1.13	4.00
B_2C_4	111	32	-28	+28	7.06	24.50
B_3C_1	111	32	-19	+19	3.25	11.28
B_3C_2	430	122	-27	+27	1.70	59.75
B_3C_3	222	75	-30	+30	3.38	12.00
B_3C_4	215	60	-33	+33	5.06	18.15

X^2 (11)= 294.54 Sig.<.001

TABLE 5.6

Relative Accessibility Interaction with Relative Location in Urban Field (1960) and Change in Total Number of Manufacturing Establishments with more than 99 Employees 1952 - 1962.

Interaction	Decline		Growth		Total
B_1C_1	220	(94.02)	14	(5.98)	234
B_1C_2	133	(94.33)	8	(5.67)	141
B_1C_3	35	(74.47)	12	(25.53)	47
B_1C_4	29	(55.77)	23	(44.23)	52
B_2C_1	183	(92.42)	15	(7.58)	198
B_2C_2	368	(91.32)	35	(8.68)	403
B_2C_3	128	(86.49)	20	(13.51)	148
B_2C_4	72	(60.00)	48	(40.00)	120
B_3C_1	156	(69.33)	69	(30.67)	225
B_3C_2	548	(74.15)	191	(25.85)	739
B_3C_3	327	(72.35)	125	(27.65)	452
B_3C_4	218	(63.56)	125	(36.44)	343
Total	2417	(77.92)	685	(22.08)	3102

	Expected Frequencies		Deviations From Observed Frequencies		Chi-Square Components	
	Decline	Growth	Decline	Growth	Decline	Growth
B_1C_1	182	52	+38	-38	7.93	27.77
B_1C_2	110	31	+23	-23	4.81	17.06
B_1C_3	37	10	- 2	+ 2	.11	.40
B_1C_4	40	12	-11	+11	3.02	10.08
B_2C_1	154	44	+29	-29	5.46	19.11
B_2C_2	314	89	+54	-54	9.29	32.76
B_2C_3	115	33	+13	-13	1.47	5.12
B_2C_4	94	26	-22	+22	5.15	18.62
B_3C_1	175	50	-19	+19	2.06	7.22
B_3C_2	576	163	-28	+28	1.36	4.81
B_3C_3	352	100	-25	+25	1.78	6.25
B_3C_4	268	75	-50	+50	9.33	33.33

X^2 (11)= 234.30 Sig.<.001

TABLE 5.6

Interaction of Change in Relative Accessibility (1950-1960) with Relative Location in Urban Field (1960) and Change in Total Number of Manufacturing Establishments with more than 99 Employees 1952 - 1962.

Interaction	Manufacturing Change: Decline		Manufacturing Change: Growth		Total
b_1C_1	445	(83.18)	90	(16.82)	535
b_1C_2	972	(81.14)	226	(18.86)	1198
b_1C_3	461	(76.45)	142	(23.55)	603
b_1C_4	287	(66.59	144	(33.41)	431
b_2C_1	114	(93.44)	8	(6.56)	122
b_2C_2	77	(90.59)	8	(9.41)	85
b_2C_3	29	(65.91)	15	(34.09)	44
b_2C_4	32	(38.10)	52	(61.90)	84
Total	2417	(77.92)	685	(22.08)	3102

	Expected Frequencies		Deviations From Observed Frequencies		Chi-Square Components	
	Decline	Growth	Decline	Growth	Decline	Growth
b_1C_1	417	118	+28	-28	1.88	6.64
b_1C_2	933	265	+39	-39	1.63	5.74
b_1C_3	470	133	- 9	+ 9	.17	.61
b_1C_4	335	96	-48	+48	6.88	24.00
b_2C_1	95	27	+19	-19	3.80	13.37
b_2C_2	66	19	+11	-11	1.83	6.37
b_2C_3	34	10	- 5	+ 5	.74	2.50
b_2C_4	67	17	-35	+35	18.28	72.06

X^2 (7)= 166.50 Sig.<.001

proportion of 22.08 per cent. This compares with 18.81 per cent and 20.35 per cent respectively in 1950 and 1960 for B_1C and 22.31 and 17.44 for the B_2C.

To summarize, given the chi-square deviations in Table 5.6, the interaction between dispersion and decentralization factors is systematic and clear: being close to a FEA city (C_4) enhances the chances of experiencing growth, given low or medium population potential. Such proximity is less important for counties having high levels of population potential. In other words, as accessibility increases, proximity to urban centers is not highly significant, thus implying decentralization to some degree even for high investment such as is involved in new large establishments.

Relative accessibility change and relative location in urban fields also shows systematic relationships. Only three groups have positive deviations that are highly significant: b_1C_4, b_2C_1 and b_2C_4. These positive deviations imply that proximity to urban centers is growth-attracting and it is more so if it is accompanied by rapid change in relative accessibility. Second, rapid relative accessibility change can overcome peripheral location in urban fields, dispersion thus implying decentralization.

Relative Location and Change Differentials

The first step in the analysis of variance, that of an arbitrarily specified order of effects in 1950 (Table 5.7) indicates that four factors are significantly associated with the variations in manufacturing rates of change. These are: FEA city size, county population potential, level of daily commuting from county to FEA city, and the interaction between relative accessibility and relative location in urban fields.

For the present purposes, it is the third sub-table of Table 5.7 which is the most important. It shows that of the four effects asso-

TABLE 5.7

Analysis of Variance 1950

Effect	Mean Square	Degrees of Freedom	F. Ratio	Significance
A	38.12	2 ,3066	27.03	<.0001
B	44.66	2 ,3066	31.66	<.0001
C	7.39	3 ,3066	5.24	<.0014
AB	1.79	4 ,3066	1.27	<.9594
AC	1.42	6 ,3066	1.00	<.4170
BC	4.28	6 ,3066	3.04	<.0058
ABC	.67	12 ,3066	.48	<.9291

Dependent variable variance is 1.41 with 3066 degrees of freedom.

ANOVA: Each Effect Alone

Effect	Mean Square	Degrees of Freedom	F. Ratio	Significance
A	38.12	2 ,3006	27.03	<.0001
B	72.31	2 ,3066	51.26	<.0001
C	21.77	3 ,3066	15.44	<.0001
AB	3.87	4 ,3066	2.75	<.0268
AC	5.13	6 ,3066	3.64	<.0014
BC	8.40	6 ,3066	5.95	<.0001
ABC	5.72	12 ,3066	4.05	<.0001

ANOVA: Significant Effects Only

Effect	Mean Square	Degrees of Freedom	F. Ratio	Significance
B	72.21	2 ,3066	51.26	<.0001
A	10.47	2 ,3066	7.42	<.0007
C	7.39	3,,3066	5.24	<.0014
BC	4.70	6 ,3006	3.33	<.0029
C	21.77	3 ,3066	15.44	<.0001
B	54.58	2 ,3066	38.69	<.0001
A	6.63	2 ,3066	4.70	<.0092
BC	4.70	6 ,3066	3.33	<.0029

Least Square Estimates of Effects

General Mean	+.71
B_1	-.35
B_2	-.25
A_1	-.20
A_2	.00
C_1	-.25
C_2	-.34
C_3	-.26
B_1C_1	-.57
B_1C_2	-.42
B_1C_3	-.57
B_2C_1	-.40
B_2C_2	-.25
B_2C_3	-.41

ciated with the change in total number of manufacturing establishments, the most important is relative accessibility. That is, in order to attract growth of large manufacturing establishments, high levels of relative accessibility were, for the period of the fifties, a necessary prerequisite. This factor alone contributes more to the explanation of the variance than all other factors put together. The three other factors describe those conditions that help attract industrial growth once high levels of accessibility are assured.

Analysis of variance for the year 1960 confirms the previous conclusions (Table 5.8). The general attainment of higher levels of relative location by 1960 did not result in a change in the relative importance of locational factors attracting new large manufacturing establishments. The main change appears to be somewhat of a decline in the importance of the interaction between relative accessibility and relative location within the economic area. The conclusion from the above is to confirm the importance of continued growth, which as a result of relative location shifts and of hierarchical diffusion, dispersion and decentralization, diminishes the extent of the periphery and extends the number of core areas. This occurs in spite of the fact that the locational attributes of counties that are necessary to attract large scale industry do not change substantially through time.

As might have been predicted with the above conclusions in hand, change in the relative location components of counties during the 1950's are not strongly associated with the change in the number of large manufacturing establishments (Table 5.9). Even the two factors that ultimately emerge as significant are not overly powerful (standard errors are large compared to the least square estimates). Perhaps the most interesting conclusion is that rapid population change is <u>negatively</u> associated with the change in the number of large plants (a_4 has a negative least square estimate). Further, in areas that

TABLE 5.8

Analysis of Variance--1960

Effect	Mean Square	Degrees of Freedom	F. Ratio	Significance
A	38.15	2 ,3066	27.08	<.0001
B	48.92	2 ,3066	34.72	<.0001
C	7.06	3 ,3066	5.01	<.0019
AB	2.20	4 ,3066	1.56	<.1801
AC	1.08	6 ,3066	.77	<.5988
BC	3.78	6 ,3066	2.68	<.1711
ABC	.77	12 ,3066	.54	<.7220
ANOVA: Each Effect Alone				
A	38.15	2 ,3066	27.08	<.0001
B	72.89	2 ,3066	51.74	<.0001
C	21.77	3 ,3066	15.45	<.0001
AB	1.06	4 ,3066	.76	<.5567
AC	5.30	6 ,3066	3.76	<.0010
BC	5.46	6 ,3066	3.88	<.0008
ABC	4.64	12 ,3066	3.29	<.0001
ANOVA: Significant Effects Only				
B	72.89	2 ,3066	51.73	<.0001
A	14.18	2 ,3066	10.07	<.0001
C	7.06	3 ,3066	5.01	<.0019
BC	3.79	6 ,3006	2.69	<.0131

Least Square Estimates of Effects

General Mean	+.66
B_1	-.27
B_2	-.33
A_1	-.23
A_2	+.01
C_1	-.29
C_2	-.38
C_3	-.28
B_1C_1	-.41
B_1C_2	-.51
B_1C_3	-.48
B_2C_1	-.48
B_2C_2	-.15
B_2C_3	-.32

TABLE 5.9

Analysis of Variance (1950-1960)

Effect	Mean Square	Degrees of Freedom	F. Ratio	Significance
a	22.23	3 ,3074	15.38	<.0001
b	.53	1 ,3074	.37	<.5464
C	8.60	3 ,3074	5.97	<.0005
ab	.88	3 ,3074	.61	<.6112
aC	1.78	9 ,3074	1.23	<.2713
bC	1.49	3 ,3074	1.03	<.3739
abC	N.A.	N.A.	N.A.	N.A.
ANOVA: Each Effect Alone				
a	22.23	3 ,3074	15.38	<.0001
b	.45	1 ,3074	.31	<.5765
C	21.77	3 ,3074	15.06	<.0001
ab	10.37	3 ,3074	7.17	<.0001
aC	12.05	9 ,3074	8.33	<.0001
bC	4.51	3 ,3074	3.12	<.0248
abC	3.09	9 ,3074	2.14	<.0235
ANOVA: Significant Effects Only				
a	22.23	3 ,3074	15 38	<.0001
aC	4.75	9 ,3074	3.28	<.0006

Least Square Estimates of Effects

General Mean	+ .89
a_1	+ .15
a_2	+ .07
a_3	+ .06
a_1c_1	- .41
a_1c_2	- .23
a_1c_3	+ .16
a_2c_1	+1.12
a_2c_2	+ .16
a_2c_3	+ .28
a_3c_1	+ .81
a_3c_2	- .001
a_3c_3	- .01

did not experience FEA city population change, proximity to the central city exerts a positive effect on change. Where there was a decline in FEA city populations, closeness to the FEA city exerted a negative effect, the same applying to those counties whose FEA city increased its population by less than 20 per cent (a_3C_4 has a negative least square estimate). Such results are extremely difficult to interpret unless more is known about population change and its regional distribution, about which more will be said later.

The analysis of variance relating initial relative location of counties and change in relative location to the change in the total number of large manufacturing establishments is presented in Table 5.10. One can see that given the prescribed order, three effects emerge as significant: FEA city population in 1950 with the change in its population between 1950 and 1960, county population potential in 1950 and its change between 1950 and 1960, and the relative location of counties in their economic area. All other effects are insignificant, and Aa and C are secondary. When the order of entry of variables into the analysis is modified to: Bb, Aa, C or to Bb, C, Aa, the initial population of the FEA city and its change become insignificant, resulting in the conclusion that what is most strongly associated with the change in the number of large manufacturing establishments is initial level of relative accessibility and the change in such accessibility, and relative location within the urban field, as follows: B_1b_1, B_2b_1, and B_3b_2 exerted negative influence on growth, and B_1b_2, B_2b_2 exerted positive influence; C_1, C_2, C_3 exerted negative influences and C_4 a positive one. In other words, having a low or medium initial population potential and experiencing only average improvements in relative accessibility, or having a high initial relative accessibility and experiencing fast growth of such accessibility resulted in lower change in the number of large manufacturing establishments than in those counties that had low or medium initial

TABLE 5.10

Analysis of Variance 1950 and 1950-1960

Effect	Mean Square	Degrees of Freedom	F. Ratio	Significance
Aa	6.24	6 ,2959	4.33	<.0003
Bb	52.52	2 ,2959	36.45	<.0001
C	8.73	3 ,2959	6.06	<.0005
AaBb	1.15	12 ,2959	.80	<.6544
AaC	1.95	18 ,2959	1.35	<.1449
BbC	2.35	6 ,2959	1.63	<.1339
ANOVA: Each Effect Alone				
Aa	6.24	6 ,2959	4.33	<.0003
Bb	63.45	2 ,2959	44.04	<.0001
C	21.77	3 ,2959	15.11	<.0001
AaBb	3.99	12 ,2959	2.77	<.0010
AaC	4.96	18 ,2959	3.44	<.0001
BbC	4.03	6 ,2959	2.80	<.0103
ANOVA: Significant Effects Only.				
Bb	63.45	2 ,2959	44.04	<.0001
C	12.42	3 ,2959	8.62	<.0001

Least Square Estimates of Effects

General Mean	+.74
B_1b_1	-.94
B_2b_1	-.66
C_1	-.03
C_2	-.37
C_3	-.30

relative accessibility levels that changed rapidly or high initial accessibility but little change. Finally, given the change in relative accessibility, location in other than the immediate suburbanization zone of the FEA modified the accessibility effect in a negative direction. These results imply that the two crucial processes explaining the spatial distribution of growth or decline in the number of large manufacturing establishments are dispersion and suburbanization.

These results do not contradict the findings in the previous sections. Hierarchical diffusion and decentralization beyond the suburbanization zones did occur, but were not as significant as one might think. Indeed, what the analysis demonstrates is that when the initial status and dynamic changes are taken into account simultaneously, it is the process of dispersion that dominated the scene during the 1950's, at least with respect to the change in the number of large manufacturing plants.

CHANGE IN NUMBERS OF MEDIUM-SIZED MANUFACTURING ESTABLISHMENTS

The locational determinants of medium-size manufacturing establishments could be anticipated to fall somewhere in between those of large plants and those of total manufacturing establishments discussed already. Small manufacturing plants (with less than 20 employees) have a locational pattern not unlike the distribution of population, being present anywhere in the country where a sufficient market and an adequate labor force exist. Moreover, as the analysis of the change in total number of manufacturing establishments already has indicated, we are approaching a situation where the population size of urban centers of economic areas is not an important factor anymore. This fact, in view of the general decline in the number of large and medium size plants, leads us to attribute the growth of all manufacturing plants to the proliferation of small plants. (The mean change per county in the number of all manufacturing plants in the 1952-1962 period was 53.8 per cent, in the number of medium size plants it was -5.5 per cent, and in the number of large plants -30.9 per cent). Growth of small-scale industry was seen to be widely diffused at all levels of the hierarchy, but change in the number of large manufacturing plants was still significantly related to city size and the economies associated with it. Yet large plants also showed a strong decentralization trend, which indicates the importance of site requirements of such plants and a growing locational flexibility within urban fields. Medium-size plants require a larger market than small ones, are more dependent on inter-industry linkages and face-to-face contacts with customers than large plants, are less sensitive than large plants to site requirements, and probably also more dependent than large plants on external and urbanization economies. For these reasons, one might expect the relative importance of the spatial determinants of change in the number of medium-size manufacturing establishments (20 to 99 employees) to differ from those of all establishments or large establishments.

Hierarchical Diffusion

Previously, hierarchical diffusion was observed with respect to the change of all manufacturing establishments and also with respect to large manufacturing establishments, but at a much lower rate. In the former case three out of five counties associated with small FEA cities experienced growth between 1952 and 1962, but in the latter case only one out of eight counties grew. The ratio with respect to medium size plants, as seen in Table 5.11, falls in between -- approximately one out of four counties in both 1950 and 1960. One concludes that hierarchical diffusion is most extensive for small-scale plants, and least extensive for large plants. Thus, the city-size effect on the location of new industrial plants approaches insignificance for

TABLE 5.11

Urban Hierarchy (1950) and Change in Total Number of Manufacturing Establishments with 20 to 99 Employees 1952 - 1962.

Manufacturing Change	A_1		A_2		A_3		Total	P
Decline	808	(75.66)	629	(61.73)	564	(55.57)	2001	(64.51)
Growth	260	(24.34)	390	(38.27)	451	(44.43)	1101	(35.49)
Total	1068		1019		1015		3102	

	Expected Frequencies			Deviations From Observed Frequencies			Chi-Square Components		
	A_1	A_2	A_3	A_1	A_2	A_3	A_1	A_2	A_3
Decline	689	657	655	+119	-28	-91	20.55	1.19	12.64
Growth	379	362	360	-119	+28	+91	37.36	2.17	23.00

X^2 (2)= 96.91 Sig.<.001

Urban Hierarchy (1960) and Change in the Total Number of Manufacturing Establishments with 20 to 99 Employees 1952 - 1962.

Manufacturing Change	A_1		A_2		A_3		Total	P
Decline	731	(77.19)	615	(61.93)	655	(56.37)	2001	(64.51)
Growth	216	(22.81)	378	(38.07)	507	(43.63)	1101	(35.49)
Total	947		993		1162		3102	

	Expected Frequencies			Deviations From Observed Frequencies			Chi-Square Components		
	A_1	A_2	A_3	A_1	A_2	A_3	A_1	A_2	A_3
Decline	611	640	750	+120	-25	-95	23.57	.98	12.03
Growth	336	353	412	-120	+25	+95	42.86	1.77	21.90

X^2 (2)= 103.11 Sig.<.001

Population Change in the Urban Hierarchy 1950 - 1960 and Change in the Total Number of Manufacturing Establishments with more than 99 Employees 1952 - 1962.

Manufacturing Change	a_1+a_2		a_3		a_4		Total	P
Decline	825	(68.86)	554	(63.68)	622	(60.15)	2001	(64.51)
Growth	373	(31.14)	316	(36.32)	412	(39.85)	1101	(35.49)
Total	1198		870		1034		3102	

	Expected Frequencies			Deviations From Observed Frequencies			Chi-Square Components		
	a_1+a_2	a_3	a_4	a_1+a_2	a_3	a_4	a_1+a_2	a_3	a_4
Decline	773	561	667	+52	-7	-45	3.50	.09	3.04
Growth	425	309	367	-52	+7	+45	6.36	.16	5.52

X^2 (2)= 18.67 Sig. .001

small plants, is important for medium-size plants and it is most significant for the location of new large plants. On the other hand, growth in FEA city population is not strong, whichever variable indicating change in the total number of manufacturing establishments is considered, being only slightly in favor of counties associated with large urban centers. The reduction of the periphery between 1950 and 1960 was the result of the general shift of the urban hierarchy upwards, and this change resulted in increased relative industrial concentration of growing counties in the major urban regions of the country in 1960 as compared to 1950:

	Number of Growing Counties	Per cent of all Counties	Per cent in					
			A_1		A_2		A_3	
			1950	1960	1950	1960	1950	1960
All Plants	2233	71.99	30.04	26.11	33.86	33.72	36.09	40.17
Medium Size Plants	1101	35.49	23.61	19.62	35.42	34.33	40.96	46.05
Large Size Plants	685	22.08	21.31	16.64	37.52	36.50	41.17	46.86

The ironical result is that continuous economic growth and substantia diffusion increases the concentration of economic activity in the largest urban regions. However, the number of such regions is growing, and the periphery is thereby reduced. Change in FEA city populations and change in total number of medium-size manufacturing establishments are significantly related: counties associated with urban centers which experienced decline or did not change their population during the 1950's were more likely to experience decline in number of medium-size manufacturing establishments; counties associated with urban centers growing only modestly showed similar trends, but the deviation from the national proportion is very small; finally, counties in urban fields where the central city grew very rapidly had much greater growth probabilities.

Dispersion

There was substantial dispersion of medium-size manufacturing establishments, whether it is related to relative accessibility in 1950 or in 1960 (see Table 5.12). The deviation pattern is by now a familiar one. The groups of counties having low or medium population potential deviated negatively from the national proportions of growth and decline and those having high population potential deviated positively. Group B_1 deviations became weaker by 1960, while those of group B_2 shifted in direction and those of B_3 became stronger. Thus, by 1960, 67 per cent of all growing counties were in areas of high relative accessibility as compared to only 52 per cent in 1950. This is clearly the result of an upward shift of relative accessibility of the entire system.

Decentralization in Urban Fields

Decentralization of industrial change is apparent (Table 5.13). One out of five counties in the periphery experienced growth. This ratio increases to 3 of 5 counties closest to urban centers. The deviation pattern is consistent: closer to the urban center the deviation changes from negative to positive, and the positive deviations are very high in the suburbanization zone. However, a marked geographical decentralization took place within urban fields during the 1950's.

Hierarchical Diffusion and Dispersion

There is a statistically-significant relationship between hierarchical status and dispersion (Table 5.14). However, 92 per cent of the total chi-square value is derived from only three groups: A_1B_1, A_2B_3, and A_3B_3. That is, given the relative status and accessibility in 1950, only in these three groups are the deviations from expectations significant. The direction of deviations shows that location in the periphery in 1950 (A_1B_1) exerted a highly negative

TABLE 5.12

Relative Accessibility (1950) in the Total Number of Manufacturing Establishments with 20 - 99 Employees 1952 - 1962.

Manufacturing Change	B_1	B_2	B_3	Total	P
Decline	589 (76.88)	677 (66.05)	735 (56.06)	2001	(64.51)
Growth	177 (23.11)	348 (33.95)	576 (43.94)	1101	(35.49)
Total	766	1025	1311	3102	

	Expected Frequencies			Deviations From Observed Frequencies			Chi-Square Components		
	B_1	B_2	B_3	B_1	B_2	B_3	B_1	B_2	B_3
Decline	494	661	846	+95	+16	-111	18.37	.39	14.56
Growth	272	364	465	-95	-16	+111	33.8	.70	6.50

X^2 (2)= 93.60 Sig.< .001

Relative Accessibility (1960) and Change in Total Number of Manufacturing Establishments with 20 - 99 Employees 1952 - 1962.

Manufacturing Change	B_1	B_2	B_3	Total	P
Decline	360 (75.95)	619 (71.23)	1022 (58.10)	2001	(64.51)
Growth	114 (24.05)	250 (28.77)	737 (41.90)	1101	(35.49)
Total	474	869	1759	3102	

	Expected Frequencies			Deviations From Observed Frequencies			Chi-Square Components		
	B_1	B_2	B_3	B_1	B_2	B_3	B_1	B_2	B_3
Decline	306	561	1134	+54	+58	-112	9.53	6.00	11.06
Growth	168	308	625	-54	-58	+112	17.36	10.92	20.07

X^2 (2)= 74.94 Sig.< .001

Change in Relative Accessibility (1950 - 1960) and Change in Total Number of Manufacturing Establishments with 20 - 99 Employees.

Manufacturing Change	b_1	b_2	Total	P
Decline	1804 (65.20)	197 (58.81)	2001	(64.51)
Growth	963 (34.80)	138 (41.19)	1101	(35.49)
Total	2767	335	3102	

	Expected Frequencies		Deviations From Observed Frequencies		Chi-Square Components	
	b_1	b_2	b_1	b_2	b_1	b_2
Decline	1785	216	+19	-19	.20	1.67
Growth	982	119	-19	+19	.37	3.03

X^2 (1)= 5.27 Sig.< .020

TABLE 5.13

Relative Location in Urban Fields (1960) and Change in Total Number of Manufacturing Establishments with 20-99 Employees.

Manufacturing Change

	C_1		C_2		C_3		C_4		Total	P
Decline	516	(78.54)	866	(67.50)	404	(62.44)	215	(41.75)	2001	(64.51)
Growth	141	(21.46)	417	(32.50)	243	(37.56)	300	(58.25)	1101	(35.49)
Total	657		1283		647		515		3102	

	Expected Frequencies				Deviations From Observed Frequencies				Chi-Square Components			
	C_1	C_2	C_3	C_4	C_1	C_2	C_3	C_4	C_1	C_2	C_3	C_4
Decline	424	828	417	332	+92	+38	-13	-117	19.96	1.74	.40	41.23
Growth	233	455	230	183	-92	-38	+13	+117	36.33	3.17	.73	74.80

X^2 (3) = 178.36 Sig. < .001

TABLE 5.14

Urban Hierarchy Interaction with Relative Accessibility (1960) and Change in Total Number of Manufacturing Establishments with 20-99 Employees 1952 - 1962.

Interaction	Manufacturing Change: Decline		Manufacturing Change: Growth		Total
A_1B_1	240	(82.47)	51	(17.53)	291
A_1B_2	283	(83.73)	55	(16.27)	338
A_1B_3	208	(65.41)	110	(34.59)	318
A_2B_1	81	(72.32)	31	(27.68)	112
A_2B_2	153	(63.49)	88	(36.51)	241
A_2B_3	381	(59.53)	259	(40.47)	640
A_3B_1	39	(54.93)	32	(45.07)	71
A_3B_2	183	(63.10)	107	(36.90)	290
A_3B_3	433	(54.06)	368	(45.94)	801
Total	2001	(64.51)	1101	(35.49)	3102

	Expected Frequencies		Deviations From Observed Frequencies		Chi-Square Components	
	Decline	Growth	Decline	Growth	Decline	Growth
A_1B_1	188	103	+52	-52	14.38	26.25
A_1B_2	218	120	+65	-65	19.38	35.21
A_1B_3	205	113	+ 3	- 3	.04	.08
A_2B_1	72	40	+ 9	- 9	1.12	2.02
A_2B_2	155	86	- 2	+ 2	.03	.05
A_2B_3	413	227	-32	+32	2.48	4.51
A_3B_1	46	25	- 7	+ 7	1.06	1.96
A_3B_2	187	113	- 4	+ 4	.09	.14
A_3B_3	517	284	-84	+84	13.65	24.84

X^2 (8)= 147.29 Sig.< .001

Interaction of Change in F.E.A. City Size with Change in Relative Accessibility (1950 - 1960) and Change in Total Number of Manufacturing Establishments with 20-99 Employees 1952 - 1962.

Interaction	Manufacturing Change: Decline		Manufacturing Change: Growth		Total
a_1b_1	391	(79.63)	100	(20.07)	491
a_1b_2	97	(84.35)	18	(15.65)	115
a_2b_1	330	(58.00)	239	(42.00)	569
a_2b_2	7	(30.43)	16	(69.57)	23
a_3b_1	521	(63.93)	294	(36.07)	815
a_3b_2	33	(60.00)	22	(40.00)	55
a_4b_1	562	(63.00)	330	(37.00)	892
a_4b_2	60	(42.25)	82	(57.75)	142
Total	2001	(64.51)	1101	(35.49)	3102

	Expected Frequencies		Deviations From Observed Frequencies		Chi-Square Components	
	Decline	Growth	Decline	Growth	Decline	Growth
a_1b_1	317	174	+74	-74	17.27	31.47
a_1b_2	4	41	+23	-23	7.15	12.90
a_2b_1	367	202	-37	+37	3.73	6.78
a_2b_2	15	8	- 8	+ 8	4.27	8.00
a_3b_1	525	290	- 4	+ 4	.03	.06
a_3b_2	35	20	- 2	+ 2	.11	.20
a_4b_1	575	317	+13	-13	.29	.53
a_4b_2	93	49	-31	+31	10.33	19.61

X^2 (7)= 122.73 Sig.< .001

TABLE 5.14

Urban Hierarchy Interaction with Relative Accessibility (1950) and Change in Total Number of Manufacturing Establishments with 20-99 Employees 1952 - 1962.

Interaction	Manufacturing Change				
	Decline		Growth		Total
A_1B_1	412	(83.91)	79	(16.09)	491
A_1B_2	246	(70.09)	105	(29.92)	351
A_1B_3	150	(66.37)	76	(33.63)	226
A_2B_1	116	(68.64)	53	(31.36)	169
A_2B_2	246	(66.85)	122	(33.15)	368
A_2B_3	267	(55.39)	215	(44.61)	482
A_3B_1	61	(57.55)	45	(42.45)	106
A_3B_2	185	(60.46)	121	(39.54)	306
A_3B_3	318	(52.74)	285	(47.26)	603
Total	2001	(64.51)	1101	(35.49)	3102

	Expected Frequencies		Deviations From Observed Frequencies		Chi-Square Components	
	Decline	Growth	Decline	Growth	Decline	Growth
A_1B_1	317	174	+95	-95	28.47	51.87
A_1B_2	226	125	+20	-20	1.77	3.20
A_1B_3	146	80	+ 4	- 4	.11	.20
A_2B_1	109	60	+ 7	- 7	.45	.82
A_2B_2	237	131	+ 9	- 9	.34	.62
A_2B_3	311	171	-44	+44	6.22	11.32
A_3B_1	68	38	- 7	+ 7	.72	1.29
A_3B_2	197	109	-12	+12	.73	1.32
A_3B_3	390	213	-72	+72	13.29	24.34

X^2 (8)= 147.08 Sig. .001

influence on the proportion of growing counties. This changes as one moves to groups A_2B_3 and A_3B_3. High relative accessibility and location in the urban field of large- or medium-size cities in 1950 exerted a high positive influence on the proportion of counties experiencing growth. Such a pattern is repeated in the change of all manufacturing establishments and in the change in large plant locations. Thus, in spite of simultaneous hierarchical diffusion and dispersion, groups A_2B_3 and A_3B_3 -- the relative location cores of the country -- were preferred by newly located industrial establishments. These two groups included 38.7 per cent of all counties experiencing growth in the number of all manufacturing establishments, 48.9 per cent in the case of large manufacturing establishments, and 45.41 per cent in the case of medium-size plants. Again, as plant size increases so does the concentration of growth in the above two groups.

By 1960, the geographical extent of the peripheral group A_1B_1 was reduced. The movement of counties towards better relative locations resulted in a concentration of almost half the counties in the country in groups A_2B_3 and A_3B_3. As a result, the pattern of deviations revealed for 1950 did not change, but the high negative effect manifested by the A_1B_1 group in 1950 was split with A_1B_2 in 1960. By 1960, as a result of upward movement, there was a decreased tendency of medium-size plants to locate in the A_1B_1 and A_1B_2 groups. Relatively speaking, the periphery became worse off by 1960. Only 9.6 per cent (106) of the growing counties belonged to the A_1B_1 and A_1B_2 groups in 1960 as compared to 16.7 per cent in 1950. Enterprise located in peripheral counties which they expected to improve (or knew were improving) their relative locations.

Hierarchical Diffusion and Decentralization

The existence of hierarchical diffusion and of decentralization is clearly evident in previous tables. When these two components of relative location interact it becomes evident that the preferred growth areas are in diminishing order: A_3C_4, A_2C_4 and A_1C_4 (Table 5.15), that is, in the more accessible counties within economic areas with viable urban centers. Indeed, most of the chi-square deviations are attributed to these three groups, with the addition of the strong negative deviation attributed to the periphery of economic areas with only small central cities (A_1C_1). Thus, although some decentralization is evident in urban fields of all hierarchical levels, it is relatively much more extensive in the suburban areas surrounding major foci.

The change in FEA city population is statistically associated with growth prospects in certain commuting zones: the greater the change in the central city population and the higher the commuting levels in 1960, the greater the number of medium-size manufacturing establishments locating there between 1952 and 1962.

Dispersion and Decentralization

Simultaneous dispersion and decentralization of medium-size manufacturing plants also are evident (Table 5.16). The systematic relationships between relative location and manufacturing change are these: the closer a county to an urban center, the greater the growth prospects, for any given level of relative accessibility. High levels of relative accessibility enhance the positive influence of proximity and reduce the negative effects of remoteness. The greatest growth rates are in areas with high potentials close to urban centers and the lowest in areas of low relative accessibility and at the periphery of urban fields. Further, change in relative accessibility interacts with location in urban fields to affect the geographical distribution of industrial change. Rapid change in relative accessibility affects change positively except at the periphery of urban regions.

TABLE 5.15

Urban Hierarchy (1950) Interaction with Relative Location in Urban Field (1960) and Change in Total Number of Manufacturing Establishments with 20 - 99 Employees 1952 - 1962.

Interaction	Manufacturing Change: Decline		Growth		Total
A_1C_1	488	(80.66)	117	(19.34)	605
A_1C_2	170	(75.22)	56	(24.78)	226
A_1C_3	99	(76.74)	30	(23.26)	129
A_1C_4	51	(47.22)	57	(52.78)	108
A_2C_1	7	(41.18)	10	(58.82)	17
A_2C_2	349	(68.70)	159	(31.30)	508
A_2C_3	168	(60.87)	108	(39.13)	276
A_2C_4	105	(48.16)	113	(51.84)	218
A_3C_1	21	(60.00)	14	(40.00)	35
A_3C_2	347	(63.20)	202	(36.80)	549
A_3C_3	137	(56.61)	105	(43.49)	242
A_3C_4	59	(31.22)	130	(68.78)	189
Total	2001	(64.51)	1101	(35.49)	3102

	Expected Frequencies		Deviations From Observed Frequencies		Chi-Square Components	
	Decline	Growth	Decline	Growth	Decline	Growth
A_1C_1	390	215	+98	-98	24.63	44.67
A_1C_2	146	80	+24	-24	3.94	7.20
A_1C_3	83	46	+16	-16	3.08	5.56
A_1C_4	70	38	-19	+19	5.16	9.50
A_2C_1	11	6	- 4	+ 4	1.45	2.67
A_2C_2	328	180	+21	-21	1.34	2.45
A_2C_3	178	98	-10	+10	.56	1.02
A_2C_4	141	77	-36	+36	9.19	16.83
A_3C_1	23	12	- 2	+ 2	.17	.33
A_3C_2	354	195	- 7	+ 7	.14	.25
A_3C_3	156	86	-19	+19	2.31	4.20
A_3C_4	121	68	-62	+62	31.77	56.53

X^2 (11)= 234.95 Sig.<.001

TABLE 5.15

(cont.)

Urban Hierarchy Interaction with Relative Location in Urban Fields (1960) and Change in Total Number of Manufacturing Establishments with 20 - 99 Employees 1952 - 1962.

Interaction	Manufacturing Change: Decline		Growth		Total
A_1C_1	487	(80.76)	116	(19.24)	603
A_1C_2	129	(76.33)	40	(23.67)	169
A_1C_3	75	(76.53)	23	(23.47)	98
A_1C_4	40	(51.95)	37	(48.05)	77
A_2C_1	7	(38.89)	11	(61.11)	18
A_2C_2	333	(68.24)	155	(31.76)	488
A_2C_3	171	(63.57)	98	(36.43)	269
A_2C_4	104	(47.71)	114	(52.29)	218
A_3C_1	22	(61.11)	14	(38.89)	36
A_3C_2	404	(64.54)	222	(35.46)	626
A_3C_3	158	(56.43)	122	(43.57)	280
A_3C_4	71	(32.27)	159	(67.73)	220
Total	2001	(64.51)	1101	(35.49)	3102

	Expected Frequencies: Decline	Expected Frequencies: Growth	Deviations From Observed Frequencies: Decline	Deviations From Observed Frequencies: Growth	Chi-Square Components: Decline	Chi-Square Components: Growth
A_1C_1	389	214	+98	-98	24.69	44.88
A_1C_2	109	60	+20	-20	3.67	6.67
A_1C_3	63	35	+12	-12	2.29	4.11
A_1C_4	50	27	-10	+10	2.00	3.70
A_2C_1	12	6	- 5	+ 5	2.08	4.17
A_2C_2	315	173	+18	-18	1.03	1.87
A_2C_3	174	95	- 3	+ 3	.05	.10
A_2C_4	141	77	-37	+37	9.71	17.78
A_3C_1	23	13	- 1	+ 1	.04	.08
A_3C_2	404	222	---	---	---	---
A_3C_3	181	99	-23	+23	2.92	5.34
A_3C_4	140	80	-69	+69	34.01	59.51

X^2 (11)= 230.70 Sig.< .001

TABLE 5.15

Interaction of Change in F.E.A. City Size (1950 - 1960) with Relative Location in Urban Field (1960) and Change in Total Number of Manufacturing Establishments with 20 - 99 Employees 1952 - 1962.

Interaction	Manufacturing Change: Decline		Growth		Total
$a_1+a_2C_1+C_2$	685	(74.54)	234	(25.46)	919
$a_1+a_2C_3$	87	(58.39)	62	(41.61)	149
$a_1+a_2C_4$	53	(40.77)	77	(59.23)	130
$a_3C_1+C_2$	291	(68.15)	136	(31.85)	427
a_3C_3	167	(67.07)	82	(32.93)	249
a_3C_4	96	(49.48)	98	(50.52)	194
$a_4C_1+C_2$	406	(68.35)	188	(31.65)	594
a_4C_3	150	(60.24)	99	(39.76)	249
a_4C_4	66	(34.55)	125	(65.45)	191
Total	2001	(64.51)	1101	(35.49)	3102

	Expected Frequencies		Deviations From Observed Frequencies		Chi-Square Components	
	Decline	Growth	Decline	Growth	Decline	Growth
$a_1+a_2C_1+C_2$	593	326	+92	-92	14.27	25.96
$a_1+a_2C_3$	96	53	- 9	+ 9	.84	1.53
$a_1+a_2C_4$	84	46	-31	+31	11.44	20.84
$a_3C_1+C_2$	275	152	+16	-16	.93	1.68
a_3C_3	161	88	+ 6	- 6	.22	.41
a_3C_4	125	69	-29	+29	6.73	12.19
$a_4C_1+C_2$	383	211	+23	-23	1.38	2.51
a_4C_3	161	88	-11	+11	.75	1.38
a_4C_4	123	68	-57	+57	26.41	47.78

X^2 (7)= 177.30 Sig.<.001

TABLE 5.16

Relative Accessibility (1950) Interaction with Relative Location in Urban Field (1960) and Change in Total Number of Manufacturing Establishments with 20-99 Employees 1952 - 1962.

Interaction	Manufacturing Change: Decline		Growth		Total
B_1C_1	293	(87.46)	42	(12.54)	335
B_1C_2	196	(78.40)	54	(21.60)	250
B_1C_3	59	(70.24)	25	(29.76)	84
B_1C_4	41	(42.47)	56	(57.73)	97
B_2C_1	130	(72.62)	49	(27.38)	179
B_2C_2	325	(67.57)	156	(32.43)	481
B_2C_3	154	(69.37)	68	(30.63)	222
B_2C_4	68	(47.55)	75	(52.45)	143
B_3C_1	93	(65.03)	50	(34.97)	143
B_3C_2	345	(62.50)	207	(37.50)	552
B_3C_3	191	(56.01)	150	(43.99)	341
B_3C_4	106	(38.54)	169	(61.46)	275
Total	2001	(64.51)	1101	(35.49)	3102

	Expected Frequencies: Decline	Expected Frequencies: Growth	Deviations From Observed Frequencies: Decline	Deviations From Observed Frequencies: Growth	Chi-Square Components: Decline	Chi-Square Components: Growth
B_1C_1	119	216	+77	-77	27.45	49.82
B_1C_2	89	161	+35	-35	7.61	13.76
B_1C_3	27	54	+ 5	- 5	.46	.83
B_1C_4	34	63	-22	+22	7.68	14.24
B_2C_1	64	115	+15	-15	1.96	3.52
B_2C_2	171	310	+15	-15	.73	1.32
B_2C_3	79	143	+11	-11	.85	1.53
B_2C_4	51	92	-24	+24	6.26	11.29
B_3C_1	51	356	+ 1	- 1	.01	.02
B_3C_2	196	220	-11	+11	.34	.62
B_3C_3	121	179	-29	+29	3.82	6.95
B_3C_4	96	242	-73	+73	29.77	55.51

X^2 (11)= 246.35 Sig.< .001

TABLE 5.16
(cont.)

Relative Accessibility Interaction with Relative Location in Urban Field (1960) and Change in Total Number of Manufacturing Establishments with More Than 99 Employees 1952 - 1962.

Interaction	Manufacturing Change: Decline		Growth		Total
B_1C_1	200	(85.47)	34	(14.53)	234
B_1C_2	105	(74.47)	36	(25.43)	141
B_1C_3	31	(65.96)	16	(34.04)	47
B_1C_4	24	(46.15)	28	(53.85)	52
B_2C_1	167	(84.34)	31	(15.66)	198
B_2C_2	291	(72.21)	112	(27.79)	403
B_2C_3	108	(72.97)	40	(27.03)	148
B_2C_4	53	(44.17)	67	(55.83)	120
B_3C_1	149	(66.22)	76	(33.78)	225
B_3C_2	470	(63.60)	269	(36.40)	739
B_3C_3	265	(58.63)	187	(41.37)	452
B_3C_4	138	(40.23)	205	(59.77)	343
Total	2001	(64.51)	1101	(35.49)	3102

	Expected Frequencies		Deviations From Observed Frequencies		Chi-Square Components	
	Decline	Growth	Decline	Growth	Decline	Growth
B_1C_1	151	83	+49	-49	14.90	28.93
B_1C_2	91	50	+14	-14	2.15	3.92
B_1C_3	30	17	+ 1	- 1	.03	.06
B_1C_4	34	18	-10	+10	2.94	5.56
B_2C_1	128	70	+39	-39	11.88	21.73
B_2C_2	260	143	+31	-31	3.70	6.72
B_2C_3	95	53	+13	-13	1.78	3.19
B_2C_4	77	43	-24	+24	7.48	13.40
B_3C_1	145	80	+ 4	- 4	.11	.20
B_3C_2	477	262	- 7	+ 7	.10	.19
B_3C_3	292	160	-27	+27	2.50	4.56
B_3C_4	221	142	-83	+83	21.17	48.51

X^2 (11)= 216.71 Sig.< .001

TABLE 5.16
(cont.)

Interaction of Change in Relative Accessibility (1950 - 1960) with Relative Location in Urban Field (1960) and Change in Total Number of Manufacturing Establishments with 20-99 Employees 1952 - 1962.

Interaction	Manufacturing Change: Decline		Growth		Total
b_1C_1	416	(77.76)	119	(22.24)	535
b_1C_2	813	(67.86)	385	(32.14)	1198
b_1C_3	378	(62.69)	225	(37.31)	603
b_1C_4	197	(45.71)	234	(54.29)	431
b_2C_1	100	(81.97)	22	(18.03)	122
b_2C_2	53	(62.35)	32	(37.65)	85
b_2C_3	26	(59.09)	18	(40.91)	44
b_2C_4	18	(21.43)	66	(78.57)	84
Total	2001	(64.51)	1101	(35.49)	3102

	Expected Frequencies		Deviations From Observed Frequencies		Chi-Square Components	
	Decline	Growth	Decline	Growth	Decline	Growth
b_1C_1	345	190	+71	-71	14.61	26.53
b_1C_2	773	425	+40	-40	2.07	3.76
b_1C_3	389	214	-11	+11	.31	.56
b_1C_4	278	153	-81	+81	23.60	42.88
b_2C_1	77	45	+23	-23	6.87	11.76
b_2C_2	55	30	- 2	+ 2	.07	.13
b_2C_3	28	16	- 2	+ 2	.14	.25
b_2C_4	56	28	-38	+38	25.79	51.57

X^2 (7)= 210.90 Sig.< .001

Relative Location and Change Differentials

The analysis of variance in Table 5.17 when the order of effects is arbitrary specified from A to ABC reveals that only two factors are significant: hierarchical status and relative accessibility. This finding is confirmed by the fact that in step two, the same two effects are the only ones significant. When accessibility is controlled, the importance of city size declines substantially (alternative 2), whereas when city size is controlled accessibility is still a significant factor, at least more significant than A in alternative 2. That is, if a single relative location component is looked for as the most important one affecting the change differentials, it is relative accessibility. Low and medium potentials depress the rate of change while high levels of relative accessibility increase it. Similarly A_1 exerts a negative influence on change, A_2 weakly so, and A_3 has a positive influence.

It is surprising to see the overwhelming importance of relative accessibility with respect to change in total number of medium size plants when actual change is the basis of the analysis, and to observe the lack of significance of all interaction effects. The dominant process seems to be that of dispersion. The inference can be made that hierarchical diffusion and decentralization are more derivative than independent processes:

Significant Initial Conditions Affecting Manufacturing Change

All Plants	Medium Size	Large Size
1. BC	B	B
2. B	A (?)	A
		C
		BC

The question that arises is why different combinations of initi relative components influenced change in the number of manufacturing plants of different sizes. One answer could be that the smaller the

TABLE 5.17

Analysis of Variance--1950

Effect	Mean Square	Degrees of Freedom	F. Ratio	Significance
A	24.41	2 ,3066	5.07	<.0064
B	13.53	2 ,3066	2.81	<.0603
C	3.89	3 ,3066	.81	<.4924
AB	2.81	4 ,3066	.58	<.6762
AC	2.36	6 ,3066	.49	<.8183
BC	4.38	6 ,3066	.91	<.4897
ABC	1.15	12 ,3066	.24	<.9964

ANOVA: Each Effect Alone

Effect	Mean Square	Degrees of Freedom	F. Ratio	Significance
A	24.41	2 ,3066	5.07	<.0064
B	27.73	2 ,3066	5.76	<.0032
C	10.91	3 ,3066	2.26	<.0783
AB	1.91	4 ,3066	.40	<.8114
AC	3.65	6 ,3066	.76	<.6034
BC	4.39	6 ,3066	.91	<.4882
ABC	3.17	12 ,3066	.66	<.7934

ANOVA: Significant Effects Only

	Effect	Mean Square	Degrees of Freedom	F. Ratio	Significance
1.	A	24.41	2 ,3066	5.07	<.0064
	B	13.53	2 ,3066	2.81	<.0603
2.	B	27.73	2 ,3066	5.76	<.0032
	A	10.20	2 ,3066	2.12	<.1199
3.	B	27.73	2 ,3066	5.76	<.0032

Least Square Estimates of Effects

For Alternatives 1, 2 above:

General Mean	+.93
B_1	-.24
B_2	-.17
A_1	-.19
A_2	-.01

For Alternative 3 above:

General Mean	+.92
B_1	-.33
B_2	-.20

plant the greater the reliance on relative accessibility and the less the weight of the other factors. As plant size increases, accessibility remains significant, but it is not the only determinant. City size is irrelevant for small plants, but becomes increasingly important with increases in plant size. The significance of the C and BC effects on large plants can be interpreted as that of the increasing importance of locational differences within the largest market areas. In any event, <u>it is clear that the most important initial condition capable of predicting the change in the number of manufacturing plants during the 1950's was the relative accessibility of counties to markets at that time, lending support to the hypothesis of initial advantage and cumulative causation</u>. This fact does not preclude hierarchical diffusion, dispersion, and decentralization, but one suspects that hierarchical diffusion, dispersion and decentralization were possible primarily because of the shift of the entire system to higher levels of relative location.

The analysis of variance for 1960 is similar to that for 1950 (Table 5.18). The two main effects associated with the rate of change in the number of medium-size manufacturing establishments are again relative accessibility and the size of the urban center of the economic areas to which counties belong. In contrast to the 1950 results, however, there is no question, statistically speaking, that the two main effects are <u>equally</u> significant. However, the higher order analysis of variance presented in Table 5.20 demonstrates very clearly that initial level of relative accessibility, together with the change in such accessibility, are the most significant factors associated with the actual rate of change in the number of medium-sized manufacturing establishments. Introduction of the accessibility change variable eliminates other main effects, including the hierarchical component.

TABLE 5.18

Analysis of Variance (1960)

Effect	Mean Square	Degrees of Freedom	F. Ratio	Significance
A	24.05	2 ,3066	4.99	<.0068
B	14.90	2 ,3066	3.09	<.0452
C	3.74	3 ,3066	.78	<.5091
AB	8.27	4 ,3066	1.71	<.1421
AC	2.09	6 ,3066	.44	<.8561
BC	2.76	6 ,3066	.57	<.7541
ABC	1.78	12 ,3066	.37	<.9743

The dependent variable variance is 4.82 with 3066 degrees of freedom.

ANOVA: Each Effect Alone

Effect	Mean Square	Degrees of Freedom	F. Ratio	Significance
A	24.05	2 ,3066	5.00	<.0068
B	23.91	2 ,3066	4.97	<.0070
C	10.91	3 ,3066	2.27	<.0779
AB	8.34	4 ,3066	1.73	<.1391
AC	3.55	6 ,3066	.74	<.6213
BC	4.63	6 ,3066	.96	<.4510
ABC	4.69	12 ,3066	.98	<.4706

ANOVA: Significant Effects Only

Effect	Mean Square	Degrees of Freedom	F. Ratio	Significance
A	24.05	2 ,3066	5.00	<.0068
B	14.90	2 ,3066	3.09	<.0452
B	23.91	2 ,3066	4.97	<.0070
A	15.04	2 ,3066	3.12	<.0439

Least Square Estimates of Effects

General Mean	+.91
A_1	-.24
A_2	-.02
B_1	-.08
B_2	-.23

TABLE 5.19

Analysis of Variance (1950-1960)

Effect	Mean Square	Degrees of Freedom	F. Ratio	Significance
a	8.97	3 ,3074	1.86	<.1333
b	1.63	1 ,3074	.34	<.5632
C	3.79	3 ,3074	.78	<.5046
ab	4.16	3 ,3074	.86	<.4635
aC	.55	9 ,3074	.11	<.9994
bC	2.43	3 ,3074	.50	<.6809
abC	N.A.		N.A.	N.A.

ANOVA: Each Effect Alone

Effect	Mean Square	Degrees of Freedom	F. Ratio	Significance
a	8.97	3 ,3074	1.86	<.1333
b	.19	1 ,3074	.04	<.8416
C	10.91	3 ,3074	2.26	<.0788
ab	2.67	3 ,3074	.55	<.6488
aC	.56	9 ,3074	.12	<.9508
bC	1.47	3 ,3074	.30	<.9736
abC	4.85	9 ,3074	1.00	<.4316

ANOVA: Significant Effects Only

Effect	Mean Square	Degrees of Freedom	F. Ratio	Significance
C	10.91	3 ,3074	2.26	<.0788

Least Square Estimates of Effects

General Means	+ .95
C_1	- .34
C_2	- .18
C_3	- .16

TABLE 5.20

Analysis of Variance (1950-1960)

Effect	Mean Square	Degrees of Freedom	F. Ratio	Significance
Aa	3.53	6 ,2959	.71	<.6398
Bb	22.07	2 ,2959	4.45	<.0118
C	6.42	3 ,2959	1.29	<.2727
AaBb	2.26	12 ,2959	.45	<.9407
AaC	2.70	18 ,2959	.54	<.9379
BbC	1.75	6 ,2959	.35	<.9090

Dependent variable variance is 4.96 with 2959 degrees of freedom

ANOVA: Each Effect Alone

Effect	Mean Square	Degrees of Freedom	F. Ratio	Significance
Aa	3.53	6 ,2959	.71	<.6398
Bb	27.05	2 ,2959	5.45	<.0044
C	10.91	3 ,2959	2.20	<.0854
AaBb	1.94	12 ,2959	.39	<.9677
AaC	4.18	18 ,2959	.84	<.6515
BbC	4.91	6 ,2959	.99	<.4292

ANOVA: Significant Effects Only

Effect	Mean Square	Degrees of Freedom	F. Ratio	Significance
Bb	27.05	2 ,2959	5.45	<.0044

Least Square Estimates of Effects

General Mean	+.89
B_1b_1	-.68
B_2b_1	-.43

CHAPTER 6

LOCATIONAL CHANGES IN PAYROLLS AND IN VALUE ADDED BY MANUFACTURING

The two variables to be considered in this chapter, value added by manufacturing and manufacturing payroll are closely related. Yet, there is an important difference between them. Value added by manufacturing does not necessarily mean that it directly affects the county in which the plants are located. It is much more likely that payroll in manufacturing represents a direct contribution to the county where manufacturing takes place or to nearby counties where commuting to place of work occurs on a regular and extensive scale. One can safely assume that payroll remains for the most part in the urban field in which manufacturing takes place, thus providing a basis for analysis of regional multiplier effects. It is clearly unsafe to make the same assumption with respect to value added.

CHANGES IN MANUFACTURING PAYROLLS

Hierarchical Diffusion

Seven out of every ten counties in the United States experienced increases in manufacturing payrolls. Hierarchical diffusion was extensive, with growth filtering down to counties at all levels of the urban hierarchy. However, variations in the extent of change are observable. Of the counties associated with large FEA cities, four out of five showed increases in manufacturing payrolls, while the ratio declined to one of two counties associated with the lowest

TABLE 6.1

Urban Hierarchy (1950) and Change in Total Payroll in Manufacturing 1952 - 1962.

Manufacturing Change	A_1		A_2		A_3		Total	P
Decline	488	(45.69)	264	(25.91)	201	(19.80)	953	(30.72)
Growth	580	(54.31)	755	(74.09)	814	(80.20)	2149	(69.28)
Total	1068		1019		1015		3102	

	Expected Frequencies			Deviations From Observed Frequencies			Chi-Square Components		
	A_1	A_2	A_3	A_1	A_2	A_3	A_1	A_2	A_3
Decline	358	313	312	+160	-49	-111	78.04	7.67	39.49
Growth	740	706	703	-160	+49	+111	34.59	3.40	17.53

X^2 (2)= 180.72 Sig.<.001

Urban Hierarchy (1960) and Change in Total Payroll in Manufacturing 1952 - 1962.

Manufacturing Change	A_1		A_2		A_3		Total	P
Decline	447	(47.20)	256	(25.78)	250	(21.51)	953	(30.72)
Growth	500	(52.80)	737	(74.22)	912	(78.49)	2149	(69.28
Total	947		993		1162		3102	

	Expected Frequencies			Deviations From Observed Frequencies			Chi-Square Components		
	A_1	A_2	A_3	A_1	A_2	A_3	A_1	A_2	A_3
Decline	291	305	357	+156	-49	-107	83.63	7.87	32.07
Growth	656	688	805	-156	+49	+107	37.10	3.49	14.22

X^2 (2)= 178.38 Sig.<.001

Population Change in the Urban Hierarchy 1950-1960 and Change in Total Payroll in Manufacturing 1952 - 1962.

Manufacturing Change	a_1+a_2		a_3		a_4		Total	P
Decline	410	(34.22)	227	(26.09)	316	(30.56)	953	(30.72)
Growth	788	(65.78)	643	(73.91)	718	(69.44)	2149	(69.28)
Total	1198		870		1034		3102	

	Expected Frequencies			Deviations From Observed Frequencies			Chi-Square Components		
	a_1+a_2	a_3	a_4	a_1+a_2	a_3	a_4	a_1+a_2	a_3	a_4
Decline	368	267	318	+42	-40	-2	4.79	5.99	.01
Growth	830	603	716	-42	+40	+2	2.12	2.65	.01

X^2 (2)= 15.57 Sig.< .001

level of the urban hierarchy. Once again one cannot avoid the conclusion that filtering-down processes are spatially biased and highly selective. If increases in manufacturing payrolls represent industrialization, it was more extensive, relatively, at the core than in the periphery. The chi-square deviations show that these differences in the extent of change are statistically significant, lending support to simultaneous diffusion and continued prominence of core areas, whether based on initial 1950 conditions, or viewed from the 1960 hierarchical structure, indicating the stability of the urban system in the United States.

Interestingly, change in city population during the 1950-1960 period was not strongly associated with the decline or growth in payroll in manufacturing during the 1952-1962 period. In particular, rapid population growth cities did not have any more extensive growth than expected on the basis of national proportions.

Dispersion

All levels of relative accessibility enjoyed growth in payroll, the ratio going from about one in two for B_1 to four out of five for B_3. When 1950 relative accessibility is compared to the 1960 levels (see Table 6.2) one can see a relative decline in the extent of growth in B_2, medium population potential counties. On the other hand, rapid change in relative accessibility is associated in a statistically significant way with positive deviations from expectation.

Decentralization in Urban Fields

The familiar distance decay from core to periphery within urban fields reveals itself explicitly in the growth of manufacturing payrolls. The proportion of counties experiencing decline decreases gradually in sequence from 47.94 per cent in C_1 to 32.58 in C_2 to 23.89 in C_3 and to a mere 12.82 per cent in C_4, the areas closest to FEA centers.

TABLE 6.2

Relative Accessibility (1950) and Change in Total Payroll in Manufacturing 1952 - 1962.

Manufacturing Change	B_1		B_2		B_3		Total	P
Decline	391	(51.04)	325	(31.71)	953	(18.08)	953	(30.72)
Growth	375	(48.96)	700	(68.29)	1074	(81.92)	2149	(69.28)
Total	766		1025		1311		3102	

	Expected Frequencies			Deviations From Observed Frequencies			Chi-Square Components		
	B_1	B_2	B_3	B_1	B_2	B_3	B_1	B_2	B_3
Decline	235	315	403	+156	+10	-166	103.56	.32	68.38
Growth	531	710	908	-156	-10	+166	45.83	.14	30.35

X^2 (2)= 248.58 Sig. < .001

Relative Accessibility (1960) and Change in Total Payroll in Manufacturing 1952 - 1962.

Manufacturing Change	B_1		B_2		B_3		Total	P
Decline	225	(47.47)	380	(43.73)	348	(19.78)	953	(30.72)
Growth	249	(52.53)	489	(56.27)	1411	(80.22)	2149	(69.28)
Total	474		869		1759		3102	

	Expected Frequencies			Deviations From Observed Frequencies			Chi-Square Components		
	B_1	B_2	B_3	B_1	B_2	B_3	B_1	B_2	B_3
Decline	146	267	540	+79	+113	-192	42.75	47.82	68.27
Growth	328	602	1219	-79	-113	+192	19.03	21.21	30.24

X^2 (2) = 229.26 Sig. < .001

Change in Relative Accessibility (1950-1960) and Changes in Total Payroll in Manufacturing 1952 - 1962.

Manufacturing Change	b_1		b_2		Total	P
Decline	831	(30.03)	122	(36.42)	953	(30.72)
Growth	1936	(69.97)	213	(63.58)	2149	(69.28)
Total	2767		335		3102	

	Expected Frequencies		Deviations From Observed Frequencies		Chi-Square Components	
	b_1	b_2	b_1	b_2	b_1	b_2
Decline	850	103	-19	+19	.42	3.50
Growth	1917	232	+19	-19	.19	1.56

X^2 (1) = 5.67 Sig. < .020

TABLE 6.3

Relative Location in Urban Fields (1960) and Change in Total Payroll in Manufacturing 1952 - 1962.

Manufacturing Change

	C_1	C_2	C_3	C_4	Total	P
Decline	315 (47.94)	418 (32.58)	154 (23.80)	66 (12.82)	953	(30.72)
Growth	342 (53.06)	865 (67.42)	493 (76.20)	449 (87.18)	2149	(69.28)
Total	657	1283	647	515	3102	

	Expected Frequencies				Deviations From Observed Frequencies				Chi-Square Components			
	C_1	C_2	C_3	C_4	C_1	C_2	C_3	C_4	C_1	C_2	C_3	C_4
Decline	202	394	119	158	+113	+24	-45	-92	63.21	1.46	10.18	54.57
Growth	455	889	448	357	-113	-24	+45	+92	28.06	.65	4.52	23.71

X^2 (3)= 185.36 Sig.<.001

Hierarchical Diffusion and Dispersion

Table 6.4 shows great variation in the extent of growth and decline in different AB areas. Three categories produce 90.98 per cent of a very high chi-square value of 322.04: a very powerful negative deviation for the peripheral group A_1B_1, which could be interpreted either as a relative decline of the periphery or as due to the fact that low-paying industries continued to dominate such areas. The second and third deviations are both positive and they occur in groups A_2B_3 and A_3B_3. These indicate the significance of both initial FEA size and initial relative accessibility. From the perspective of the 1960 relative location, the negative deviation spreads to two groups A_1B_1 and A_1B_2, and the positive deviations are the same. The interaction of change in FEA city size with change in relative accessibility reveals three strong deviations from expectations. Two are negative (a_1b_1 and a_1b_2) and one is positive (a_2b_1). Thus, regardless of relative accessibility change, stagnant economic areas lacking a viable urban-economic center experienced less growth than other groups.

Hierarchical Diffusion and Decentralization

Some growth occurred at all levels of the hierarchy and at all levels of local accessibility, but as before, the location within the highest-status FEA's in the high-accessibility heartland produced greater than expected and more extensive growth whereas peripheral location was associated with concentration around the central city, and a greater decline than expected at the periphery. Joint filtering and decentralization did take place in the 1950's, but on a very highly selective basis. These effects are the same whether one considers initial 1950 status, 1960 status, or changes between 1950 and 1960 (Table 6.5).

TABLE 6.4

Urban Hierarchy Interaction with Relative Accessibility (1950) and Change in Total Payroll in Manufacturing 1952 - 1962.

Interaction	Manufacturing Change: Decline		Manufacturing Change: Growth		Total
A_1 1	163	(56.01)	128	(43.99)	291
A_1B_2	197	(58.28)	141	(41.72)	338
A_1B_3	87	(27.36)	231	(72.64)	318
A_2B_1	43	(38.39)	69	(61.61)	112
A_2B_2	84	(34.85)	157	(65.15)	241
A_2B_3	129	(20.16)	511	(79.84)	640
A_3B_1	19	(26.76)	52	(73.24)	71
A_3B_2	99	(34.14)	191	(65.86)	290
A_3B_3	132	(16.48)	669	(83.52)	801
Total	953	(30.72)	2149	(69.28)	3102

	Expected Frequencies		Deviations From Observed Frequencies		Chi-Square Components	
	Decline	Growth	Decline	Growth	Decline	Growth
A_1B_1	89	202	+74	-74	61.53	27.11
A_1B_2	104	234	+93	-93	83.16	36.96
A_1B_3	98	220	-11	+11	1.23	.55
A_2B_1	34	78	+ 9	- 9	2.38	1.04
A_2B_2	74	167	+10	-10	1.35	.60
A_2B_3	197	443	-68	+68	23.47	10.44
A_3B_1	22	49	- 3	+ 3	.41	.18
A_3B_2	89	201	+10	-10	1.12	.50
A_3B_3	246	555	-114	+114	52.83	23.42

X^2 (8)= 328.28 Sig.< .001

Interaction of Change in F.E.A. City Size with Change in Relative Accessibility (1950-1960) and Change in Total Payroll in Manufacturing.

Interaction	Manufacturing Change: Decline		Manufacturing Change: Growth		Total
a_1b_1	242	(49.29)	249	(50.71)	491
a_1b_2	68	(59.13)	47	(40.87)	115
a_2b_1	98	(17.22)	471	(82.78)	569
a_2b_2	2	(8.70)	21	(91.30)	23
a_3b_1	211	(25.89)	604	(74.11)	815
a_3b_2	16	(29.09)	39	(70.81)	55
a_4b_1	280	(31.39)	612	(68.61)	892
a_4b_2	36	(25.35)	106	(74.65)	142
Total	953	(30.72)	2149	(69.28)	3102

	Expected Frequencies		Deviations From Observed Frequencies		Chi-Square Components	
	Decline	Growth	Decline	Growth	Decline	Growth
a_1b_1	151	340	+91	-91	54.84	24.36
a_1b_2	35	80	+33	-33	31.11	13.61
a_2b_1	175	394	-77	+77	33.88	15.05
a_2b_2	7	16	- 5	+ 5	3.57	1.56
a_3b_1	250	565	-39	+39	6.08	2.69
a_3b_2	17	38	- 1	+ 1	.06	.03
a_4b_1	274	618	+ 6	- 6	.13	.06
a_4b_2	44	98	- 8	+ 8	1.45	.65

X^2 (6)= 189.13 Sig.< .001

TABLE 6.4
(cont.)

Urban Hierarchy Interaction with Relative Accessibility (1950) and the Change in Total Payroll in Manufacturing 1952 - 1962.

Manufacturing Change

Interaction	Decline		Growth		
A_1B_1	292	(59.47)	199	(40.53)	491
A_1B_2	140	(39.89)	211	(60.11)	351
A_1B_3	56	(24.78)	170	(75.22)	226
A_2B_1	69	(40.83)	100	(59.17)	169
A_2B_2	102	(27.72)	266	(72.28)	368
A_2B_3	93	(19.30)	389	(80.70)	482
A_3B_1	30	(28.30)	76	(71.70)	106
A_3B_2	83	(27.12)	223	(72.88)	306
A_3B_3	88	(14.59)	515	(85.41)	603
Total	953	(30.72)	2149	(69.28)	3102

	Expected Frequencies		Deviations From Observed Frequencies		Chi-Square Components	
	Decline	Growth	Decline	Growth	Decline	Growth
A_1B_1	151	340	+141	-141	131.66	58.47
A_1B_2	108	243	+32	-32	9.48	4.21
A_1B_3	69	157	-13	+13	2.45	1.08
A_2B_1	52	117	+17	-17	5.56	2.47
A_2B_2	113	255	-11	+11	1.07	.47
A_2B_3	148	334	-55	+55	20.44	9.06
A_3B_1	33	73	- 3	+ 3	.27	.12
A_3B_2	94	212	-11	+11	1.29	.57
A_3B_3	185	418	-97	+97	50.86	22.51

X^2 (8)= 322.04 Sig.< .001

TABLE 6.5

Urban Hierarchy (1950) Interaction with Relative Location in Urban Field (1960) and Change in Total Payroll in Manufacturing 1952-1962.

Interaction	Manufacturing Change				
	Decline		Growth		Total
A_1C_1	311	(51.40)	294	(48.60)	605
A_1C_2	118	(52.21)	108	(47.79)	226
A_1C_3	45	(34.88)	84	(65.12)	129
A_1C_4	14	(12.96)	94	(87.04)	108
A_2C_1	2	(11.76)	15	(88.24)	17
A_2C_2	172	(33.86)	336	(66.14)	508
A_2C_3	53	(19.20)	223	(80.80)	276
A_2C_4	37	(16.97)	181	(83.03)	218
A_3C_1	2	(5.71)	33	(94.29)	35
A_3C_2	128	(23.32)	421	(76.68)	549
A_3C_3	56	(23.14)	186	(76.86)	242
A_3C_4	15	(7.94)	174	(92.06)	189
Total	953	(30.72)	2149	(69.28)	3102

	Expected Frequencies		Deviations From Observed Frequencies		Chi-Square Components	
	Decline	Growth	Decline	Growth	Decline	Growth
A_1C_1	186	419	+125	-125	84.00	37.29
A_1C_2	69	157	+49	-49	34.80	15.29
A_1C_3	40	89	+ 5	- 5	.62	.28
A_1C_4	33	75	-19	+19	10.94	4.81
A_2C_1	5	12	- 3	+ 3	1.80	.75
A_2C_2	156	352	+16	-16	1.64	.73
A_2C_3	85	191	-32	+32	12.05	5.36
A_2C_4	67	151	-30	+30	13.43	5.96
A_3C_1	10	25	- 8	+ 8	6.40	2.56
A_3C_2	169	380	-41	+41	9.95	4.42
A_3C_3	74	168	-18	+18	4.38	1.92
A_3C_4	59	130	-44	+44	32.81	14.89

X^2 (9)= 307.08 Sig.$<$.001

TABLE 6.5
(cont.)

Urban Hierarchy Interaction with Relative Location in Urban Field (1960) and Change in Total Payroll in Manufacturing 1952 - 1962.

Interaction	Manufacturing Change: Decline		Growth		Total
A_1C_1	311	(51.58)	292	(48.42)	603
A_1C_2	89	(52.66)	80	(47.34)	169
A_1C_3	36	(36.73)	62	(63.27)	98
A_1C_4	11	(14.29)	66	(85.71)	77
A_2C_1	1	(5.56)	17	(94.46)	18
A_2C_2	169	(34.63)	319	(65.37)	488
A_2C_3	55	(20.45)	214	(79.55)	269
A_2C_4	31	(14.22)	187	(85.78)	218
A_3C_1	3	(8.33)	33	(91.67)	36
A_3C_2	160	(25.56)	466	(74.44)	626
A_3C_3	63	(22.50)	217	(77.50)	280
A_3C_4	24	(10.91)	196	(89.09)	220
Total	953	(30.72)	2149	(69.28)	3102

	Expected Frequencies		Deviations From Observed Frequencies		Chi-Square Components	
	Decline	Growth	Decline	Growth	Decline	Growth
A_1C_1	185	418	+126	-126	85.82	37.98
A_1C_2	52	117	+37	-37	26.33	11.70
A_1C_3	30	68	+ 6	- 6	1.20	.53
A_1C_4	24	53	-13	+13	7.04	3.19
A_2C_1	6	12	- 5	+ 5	4.17	2.08
A_2C_2	150	338	+19	-19	2.41	1.07
A_2C_3	83	186	-28	+28	9.45	4.22
A_2C_4	67	151	-36	+36	19.34	8.58
A_3C_1	11	25	- 8	+ 8	5.82	2.56
A_3C_2	192	434	-32	+32	5.33	2.36
A_3C_3	86	194	-23	+23	6.15	2.73
A_3C_4	67	153	-43	+43	27.60	12.08

X^2 (9) = 289.74 Sig. < .001

TABLE 6.5
(cont.)

Interaction of Change in F.E.A City Size (1950 - 1960) with Relative Location in Urban Field (1960) and Change in Total Payroll in Manufacturing 1952 - 1962.

Interaction	Manufacturing Change: Decline		Growth		Total
$a_1+a_2C_1+C_2$	366	(39.83)	553	(60.17)	919
$a_1+a_2C_3$	31	(20.80)	118	(79.20)	149
$a_1+a_2C_4$	13	(10.00)	117	(90.00)	130
$a_3C_1+C_2$	144	(33.72)	283	(66.28)	427
a_3C_3	58	(23.29)	191	(77.71)	249
a_3C_4	25	(12.89)	169	(87.11)	194
$a_4C_1+C_2$	222	(37.37)	372	(62.63)	594
a_4C_3	666	(26.51)	183	(73.49)	249
a_4C_4	28	(14.66)	163	(85.34)	191
Total	953	(30.72)	2149	(69.28)	3102

	Expected Frequencies		Deviations From Observed Frequencies		Chi-Square Components	
	Decline	Growth	Decline	Growth	Decline	Growth
$a_1+a_2C_1+C_2$	282	637	+84	-84	25.02	11.08
$a_1+a_2C_3$	46	103	-15	+15	4.89	2.18
$a_1+a_2C_4$	40	90	-27	+27	18.22	8.10
$a_3C_1+C_2$	131	296	+13	-13	1.29	.57
a_3C_3	76	173	-18	+18	4.26	1.87
a_3C_4	60	134	-35	+35	20.42	9.14
$a_4C_1+C_2$	182	412	+40	-40	8.79	3.88
a_4C_3	76	173	-10	+10	1.32	.58
a_4C_4	60	131	-32	+32	17.07	7.82

X^2 (8)= 146.50 Sig.<.001

Dispersion and Decentralization

In contrast with the AB and AC interactions, BC shows more systematic deviations from expectations in Table 6.6, and a wider spread of the variation into more groups. First, one should note the absolute existence of dispersion and decentralization, as shown in the first part of the table by the percentage distribution of growth and decline. Secondly, note the negative deviations for $B_1C_{1,2,3}$ and $B_2C_{1,2,3}$ and the positive deviation for B_1C_4, B_2C_4 and B_3C_i. Low potential results in negative deviations for all commuting zones except C_4 for B_1 and B_2. Also note that the significance of the negative deviations declines from B_1C_1 to B_2C_i. There is a shift to positive deviations for all B_3C_i groups, the significance of which culminates in B_3C_4, the most accessible areas of the country in 1950. Such a pattern is by now familiar. In addition, regardless of the change in population potential of counties between 1950 and 1960, the closer a county is to the focus of the economic area, the greater the likelihood of experiencing growth in payroll. The deviation b_2C_1 raises the speculation that rapid improvement in relative accessibility does attract industrial expansion, but could cause migration of the labor force just as well to the close and distant urban centers.

Relative Location and Change Differentials

The analysis of variance reveals that all main effects and two way interaction effects are statistically significant (Table 6.7). Each relative location component, and each interaction of components, is associated not only with the spatial extent of change in manufacturing payrolls but also with the intensity or degree of change. Not all components remain significant when other effects are controlled, however. The single factor that most affects the degree of change in manufacturing is relative location within the urban field. This factor is followed by the BC interaction. That is, relative

TABLE 6.6

Relative Accessibility (1950) Interaction with Relative Location in Urban Field (1960) and Change in Total Payroll in Manufacturing 1952 - 1962.

Interaction	Manufacturing Change				
	Decline		Growth		Total
B_1C_1	209	(62.39)	126	(37.61)	335
B_1C_2	139	(55.60)	111	(44.40)	250
B_1C_3	32	(38.10)	52	(61.90)	84
B_1C_4	11	(11.34)	86	(88.66)	97
B_2C_1	69	(38.55)	110	(61.45)	179
B_2C_2	159	(33.06)	322	(66.94)	481
B_2C_3	71	(31.98)	151	(68.02)	222
B_2C_4	26	(18.18)	117	(81.82)	143
B_3C_1	37	(25.87)	106	(74.13)	143
B_3C_2	120	(21.74)	432	(78.26)	552
B_3C_3	51	(14.96)	290	(85.04)	341
B_3C_4	29	(10.55)	246	(89.45)	275
Total	953	(30.72)	2149	(69.28)	3102

	Expected Frequencies		Deviations From Observed Frequencies		Chi-Square Components	
	Decline	Growth	Decline	Growth	Decline	Growth
B_1C_1	103	232	+106	-106	109.09	48.43
B_1C_2	77	173	+62	-62	49.92	22.22
B_1C_3	26	58	+ 6	- 6	1.38	.68
B_1C_4	30	67	-19	+19	12.03	5.39
B_2C_1	55	124	+14	-14	3.56	1.58
B_2C_2	148	333	+11	-11	.82	.36
B_2C_3	68	154	+ 3	- 3	.13	.06
B_2C_4	44	99	-18	+18	7.36	3.27
B_3C_1	44	99	- 7	+ 7	1.11	.49
B_3C_2	170	382	-50	+50	14.70	6.54
B_3C_3	105	236	-54	+54	27.78	12.36
B_3C_4	83	192	-54	+54	35.13	15.19

X^2 (11)= 379.58 Sig.< .001

TABLE 6.6
(cont.)

Relative Accessibility Interaction with Relative Location in Urban Field (1960) and Change in Total Payroll in Manufacturing 1952 - 1962.

Interaction	Manufacturing Change: Decline		Manufacturing Change: Growth		Total
B_1C_1	136	(58.12)	98	(41.88)	234
B_1C_2	71	(50.35)	70	(49.65)	141
B_1C_3	12	(25.53)	35	(74.47)	47
B_1C_4	6	(11.54)	46	(88.46)	52
B_2C_1	113	(57.07)	85	(42.93)	198
B_2C_2	181	(44.91)	222	(55.09)	403
B_2C_3	64	(43.24)	84	(56.76)	148
B_2C_4	22	(18.33)	98	(80.67)	120
B_3C_1	66	(29.33)	159	(70.67)	225
B_3C_2	166	(22.46)	573	(77.54)	739
B_3C_3	78	(17.26)	374	(82.74)	452
B_3C_4	38	(11.08)	305	(88.92)	343
Total	953	(30.72)	2149	(69.28)	3102

	Expected Frequencies		Deviations From Observed Frequencies		Chi-Square Components	
	Decline	Growth	Decline	Growth	Decline	Growth
B_1C_1	72	152	+64	-64	56.89	26.95
B_1C_2	43	98	+28	-28	18.23	8.00
B_1C_3	14	33	- 2	+ 2	.29	.12
B_1C_4	16	36	-10	+10	6.25	2.78
B_2C_1	61	137	+52	-52	44.33	19.74
B_2C_2	124	279	+57	-57	26.20	11.64
B_2C_3	45	103	+19	-19	8.02	3.50
B_2C_4	37	83	-15	+15	6.08	2.71
B_3C_1	69	156	- 3	+ 3	.13	.06
B_3C_2	227	512	-61	+61	16.39	7.27
B_3C_3	138	314	-60	+60	26.09	11.46
B_3C_4	107	236	-69	+69	44.50	20.17

X^2 (11) = 367.80 Sig. < .001

TABLE 6.6
(cont.)

Interaction of Change in Relative Accessibility (1950 - 1960) with Relative Location in Urban Field (1960) and Change in Total Payroll in Manufacturing 1952 - 1962.

Interaction	Manufacturing Change: Decline		Growth		Total
b_1C_1	246	(45.98)	289	(54.02)	535
b_1C_2	382	(31.59)	816	(68.11)	1198
b_1C_3	142	(23.55)	461	(76.45)	603
b_1C_4	61	(14.15)	370	(85.85)	431
b_2C_1	69	(56.56)	53	(43.44)	122
b_2C_2	35	(41.18)	50	(58.82)	85
b_2C_3	13	(29.55)	21	(70.45)	44
b_2C_4	5	(5.95)	79	(94.05)	84
Total	953	(30.72)	2149	(69.28)	3102

	Expected Frequencies		Deviations From Observed Frequencies		Chi-Square Components	
	Decline	Growth	Decline	Growth	Decline	Growth
b_1C_1	164	371	+82	-82	41.00	18.12
b_1C_2	368	830	+14	-14	.53	.24
b_1C_3	185	418	-43	+43	9.99	4.42
b_1C_4	132	299	-71	+71	38.19	16.86
b_2C_1	37	85	+32	-32	27.68	12.05
b_2C_2	26	59	+ 9	- 9	3.12	1.37
b_2C_3	14	30	- 1	+ 1	.07	.03
b_2C_4	27	57	-22	+22	17.93	8.49

X^2 (7)= 200.09 Sig.< .001

location within urban fields is highly significant, but the effect is differentiated by its interaction with national relative accessibility. The most important inference relates to the positive effects of C_4 and B_3. There is also a positive effect of A_3 and B_3 in the AB interaction, and a negative effect of A_2C_4 and A_3C_4. Probably this is a result of declining economies of scale or of external diseconomies in the core counties of the AC groups. Of course, this is so given prior control of the C, BC and AB effects.

One of the astonishing results is the relative accuracy of the estimated value of change based on the least squares equation (Table 6.8). Only in a few cases is the estimation far from the actual change values. This implies that the explanation of the spatial change in manufacturing payroll, as well as in other variables discussed earlier, could very well be based on the analysis of relative location components.

Change in relative location also was associated with the degree of change in manufacturing payroll. The sequence of factors that significantly affects payroll is as follows a, b, C and ab (change in FEA city size, change in relative accessibility, location with urban fields in 1960, and the interaction of FEA city population change with population potential change). Interestingly enough, when the order that variables are introduced into the analysis changes, only two factors remain significant, aC and b. Least square estimates are available only for alternative 1. Interestingly, one notes that by implication aC_4 always has a positive effect but $a_4C_{1,2,3}$ have a negative effect. Rapid change of the FEA city population resulted in a very rapid growth in manufacturing payroll but not so for a_4C_1, a_4C_2 and a_4C_3, an indication of the interrelations between high wages and salaries and rapid urbanization and industrialization.

Not surprisingly, then, when, initial relative location and its change are compared with the rate of manufacturing payroll growth, a

TABLE 6.7

Analysis of Variance--1950

Effect	Mean Square	Degrees of Freedom	F. Ratio	Significance
A	70.14	2 ,3066	5.39	<.0046
B	14.09	2 ,3066	1.08	<.3368
C	73.60	3 ,3066	5.66	<.0008
AB	101.23	4 ,3066	7.78	<.0001
AC	40.98	6 ,3066	3.15	<.0044
BC	78.66	6 ,3066	6.05	<.0001
ABC	20.99	12 ,3066	1.61	<.0802

Dependent variable variance is 13.00 with 3066 degrees of freedom

ANOVA: Each Effect Alone

Effect	Mean Square	Degrees of Freedom	F. Ratio	Significance
A	70.14	2 ,3066	5.39	<.0046
B	39.66	2 ,3066	3.05	<.0473
C	97.66	3 ,3066	7.51	<.0001
AB	74.05	4 ,3066	5.69	<.0002
AC	49.94	6 ,3066	3.84	<.0009
BC	57.85	6 ,3066	4.45	<.0002
ABC	18.21	12 ,3066	1.40	<.1572

ANOVA: Significant Effects Only

Effect	Mean Square	Degrees of Freedom	F. Ratio	Significance
C	97.66	3 ,3066	7.51	<.0001
BC	81.45	6 ,3066	6.26	<.0001
AB	52.86	4 ,3066	4.05	<.0028
AC	56.19	6 ,3066	4.32	<.0003
A	65.29	2 ,3066	5.02	<.0067
BC	57.85	6 ,3066	4.45	<.0002
C	144.86	3 ,3066	11.14	<.0001
AC	49.85	6 ,3066	3.83	<.0009
AB	62,20	4 ,3066	4.78	<.0008
A	65.29	2 ,3066	5.02	<.0067

Least Square Estimates of Effects

General Mean	+2.66
C_1	- .08
C_2	-1.19
C_3	-1.08
B_1C_1	-1.64
B_1C_2	+ .57
B_1C_3	-2.22
B_2C_1	- .58
B_2C_2	-3.50
B_2C_3	-1.06
A_1B_1	-2.36
A_1B_2	-1.93
A_2B_1	-1.00
A_2B_2	- .50
A_1C_1	-1.15
A_1C_2	+ .91
A_1C_3	+ .19
A_2C_1	+ .76
A_2C_2	+1.73
A_2C_3	+1.26
A_1	- .75
A_2	- .58

TABLE 6.8

Analysis of Variance--1960

Effect	Mean Square	Degrees of Freedom	F. Ratio	Significance
A	76.04	2 ,3066	5.78	<.0032
B	39.11	2 ,3066	2.97	<.0511
C	65.09	3 ,3066	4.95	<.0020
AB	60.52	4 ,3066	4.60	<.0011
AC	39.54	6 ,3066	3.01	<.0063
BC	45.62	6 ,3066	3.47	<.0021
ABC	10.80	12 ,3066	.82	<.6301

ANOVA: Each Effect Alone

Effect	Mean Square	Degrees of Freedom	F. Ratio	Significance
A	76.04	2 ,3066	5.78	<.0032
B	71.03	2 ,3066	5.40	<.0046
C	97.66	3 ,3066	7.42	<.0001
AB	42.07	4 ,3066	3.20	<.0124
AC	44.80	6 ,3066	3.41	<.0024
BC	36.10	6 ,3066	2.74	<.0116
ABC	19.48	12 ,3066	1.48	<.1229

ANOVA: Significant Effects Only

Effect	Mean Square	Degrees of Freedom	F. Ratio	Significance
C	97.66	3 ,3066	7.42	<.0001
BC	57.17	6 ,3066	4.35	<.0003
AB	49.94	4 ,3066	3.80	<.0044
AC	39.84	6 ,3066	3.03	<.0059
A	36.83	2 ,3066	2.80	<.0607

Least Square Estimates of Effects

General Mean	+2.59
C_1	- .14
C_2	-1.21
C_3	- .01
B_1C_1	-1.26
B_1C_2	-1.24
B_1C_3	-1.03
B_2C_1	-1.79
B_2C_2	-1.56
B_2C_3	-2.33
A_1B_1	- .18
A_1B_2	-1.81
A_2B_1	+ .14
A_2B_2	- .65
A_1C_1	-1.02
A_1C_2	+1.37
A_1C_3	+ .18
A_2C_1	+ .64
A_2C_2	+1.47
A_2C_3	+1.10
A_1	- .60
A_2	- .09

TABLE 6.9

Analysis of Variance (1950-1960)

Effect	Mean Square	Degrees of Freedom	F. Ratio	Significance
a	59.46	3 ,3074	4.53	<.0036
b	110.76	1 ,3074	8.44	<.0037
C	77.29	3 ,3074	5.89	<.0006
ab	57.75	3 ,3074	4.40	<.0043
aC	16.49	9 ,3074	1.25	<.2537
bC	116.77	3 ,3074	8.89	<.0001
abC	N.A.			

ANOVA: Each Effect Alone

Effect	Mean Square	Degrees of Freedom	F. Ratio	Significance
a	59.46	3 ,3074	4.53	<.0036
b	79.12	1 ,3074	6.02	<.0141
C	97.66	3 ,3074	7.44	<.0001
ab	5.42	3 ,3074	.41	<.7452
aC	57.29	9 ,3074	4.36	<.0001
bC	22.80	3 ,3074	1.74	<.1561
abC	12.76	9 ,3074	.97	<.4634

ANOVA: Significant Effects Only

	Effect	Mean Square	Degrees of Freedom	F. Ratio	Significance
1.	aC	57.29	9 ,3074	4.36	<.0001
	b	76.86	1 ,3074	5.85	<.0156

Least Square Estimates of Effects

General Mean	+2.69
a_1C_1	+ .28
a_1C_2	+ .90
a_1C_3	+ .48
a_2C_1	+2.82
a_2C_2	+ .92
a_2C_3	- .02
a_3C_1	+2.10
a_3C_2	+1.09
a_3C_3	+ .34
b_1	- .52

TABLE 6.10

Analysis of Variance 1950 and 1950-1960

	Effect	Mean Square	Degrees of Freedom	F. Ratio	Significance
	Aa	43.50	6 ,2959	3.38	<.0026
	Bb	44.31	2 ,2959	3.44	<.0322
	C	64.04	3 ,2959	4.97	<.0020
	AaBb	23.37	12 ,2959	1.82	<.0405
	AaC	19.63	18 ,2959	1.52	<.0715
	BbC	24.43	6 ,2959	1.90	<.0774
ANOVA:	Each Effect Alone				
	Aa	43.50	6 ,2959	3.38	<.0026
	Bb	69.18	2 ,2959	5.37	<.0047
	C	97.66	3 ,2959	7.59	<.0001
	AaBb	18.39	12 ,2959	1.43	<.1448
	AaC	19.49	18 ,2959	1.51	<.0751
	BbC	10.20	6 ,2959	.79	<.5762
ANOVA:	Significant Effects Only				
1.	Aa	43.50	6 ,2959	3.38	.0026
	Bb	44.31	2, 2959	3.44	.0322
	C	64.04	3 ,2959	4.97	.0020
	AaBb	23.37	12 ,2959	1.82	.0405
2.	C	97.66	3 ,2959	7.59	<.0001
	Aa	28.30	6 ,2959	2.20	<.0403
	AaC	23.26	18 ,2959	1.81	<.0195
3.	C	97.66	3 ,2959	7.59	<.0001
	Aa	28.30	6 ,2959	2.20	<.0403
	Bb	39.50	2 ,2959	3.07	<.0467
	AaC	21.19	18 ,2959	1.65	<.0419

Least Square Estimates of Effects

For Alternative 2.

General Mean	+2.76
C_1	+ .88
C_2	- .97
C_3	- .64
A_1a_1	-1.29
A_1a_2	+1.00
A_1a_3	-1.93
A_2a_1	+1.15
A_2a_2	+ .10
A_2a_3	- .52
$A_1a_1C_1$	-4.61
$A_1a_1C_2$	+3.18
$A_1a_1C_3$	-2.39
$A_1a_2C_1$	+1.25
$A_1a_2C_2$	- .97
$A_1a_2C_3$	-1.60
$A_1a_3C_1$	-4.23
$A_1a_3C_2$	+1.39
$A_1a_3C_3$	+1.28
$A_2a_1C_1$	-2.94
$A_2a_1C_2$	-1.56
$A_2a_1C_3$	-3.69
$A_2a_2C_1$	+ .36
$A_2a_2C_2$	- .23
$A_2a_2C_3$	- .51
$A_2a_3C_1$	- .20
$A_2a_3C_2$	+1.68
$A_2a_3C_3$	+1.82

number of combinations of three to four factors is significant. First is the hypothesized order of Aa, Bb, C and AaBb. Such a combination shows that the interaction between initial location and its change is indeed relevant in determining the degree of change in a manufacturing variable. The above order confirms the existence of hierarchical diffusion together or in spite of the dynamic changes in the urban hierarchy. Secondly, initial population potential and the change in potential determine the relative accessibility of counties to the national markets and therefore their attractiveness to industry and the change in payroll in manufacturing. The interaction between hierarchical status and dispersion and their change: AaBb is still significant, but not very powerful (F-value of 1.82). Finally, as was evident throughout the chi-square computations and the analysis of variance, the importance of the relative location within an urban field, or the intensity of association with the urban core as represented by commuting levels, is highly crucial. In fact, if one wants to point to one most important factor it is C. The direction of the C effect is demonstrated in Table 6.10. C_2 and C_3 exert negative influence, urbanization and external economies are insufficient to result in a positive effect on the average change. By implication, C_4, the inner suburban zone had above average change.

CHANGE IN VALUE ADDED BY MANUFACTURING

Hierarchical Diffusion

Value added by manufacturing increased in 66.44 per cent or 2061 counties in the United States from 1952 to 1962. The distribution of counties by growth or decline is cross-classified with hierarchical status of the economic centers of the FEA's within which counties are located in Table 6.11. The association is significant. The higher the hierarchical level of an FEA center, the greater the likelihood of a county located within the FEA to experience growth in value added. Such a generalization is supported by the pattern of

TABLE 6.11

Urban Hierarchy (1950) and Change in Total Value-Added by Manufacturing 1952 - 1962.

Manufacturing Change	A_1		A_2		A_3		Total	P
Decline	511	(47.85)	303	(29.74)	227	(22.36)	1041	(33.56)
Growth	557	(52.15)	716	(70.26)	788	(77.64)	2061	(66.44)
Total	1068		1019		1015		3102	

	Expected Frequencies			Deviations From Observed Frequencies			Chi-Square Components		
	A_1	A_2	A_3	A_1	A_2	A_3	A_1	A_2	A_3
Decline	358	342	341	+153	-39	-114	65.39	4.45	38.11
Growth	710	677	674	-153	+39	+114	32.97	2.25	19.28

X^2 (2)= 162.45 Sig.< .001

Urban Hierarchy (1960) and Change in Total Value-Added by Manufacturing 1952 - 1962.

Manufacturing Change	A_1		A_2		A_3		Total	P
Decline	469	(49.53)	291	(29.31)	281	(24.19)	1041	(33.56)
Growth	478	(50.47)	702	(70.69)	881	(75.81)	2061	(66.44)
Total	947		993		1162		3102	

	Expected Frequencies			Deviations From Observed Frequencies			Chi-Square Components		
	A_1	A_2	A_3	A_1	A_2	A_3	A_1	A_2	A_3
Decline	318	333	390	+151	-42	-109	71.70	5.30	30.46
Growth	629	660	772	-151	+42	+109	26.25	2.67	15.39

X^2 (2)= 161.77 Sig.< .001

Population Change in the Urban Hierarchy 1950 - 1960 and Changes in Total Value-Added by Manufacturing 1952 - 1962.

Manufacturing Change	a_1+a_2		a_3		a_4		Total	P
Decline	437	(36.48)	254	(29.20)	350	(33.85)	1041	(33.56
Growth	761	(63.52)	616	(70.80)	684	(66.15)	2061	(66.44)
Total	1198		870		1034		3102	

	Expected Frequencies			Deviations From Observed Frequencies			Chi-Square Components		
	a_1+a_2	a_3	a_4	a_1+a_2	a_3	a_4	a_1+a_2	a_3	a_4
Decline	402	292	347	+35	-38	+3	3.05	4.95	.03
Growth	796	578	687	-35	+38	-3	1.54	2.50	.01

X^2 (2)= 12.08 Sig.< .01

deviations from expected frequencies, which is large and negative for the medium level (A_2), and positive and high for the group of countie associated with the highest level of the urban hierarchy (A_3). Thus, in relative terms growth was more widespread in high-status than in low-status counties. However, the fact that one county out of two in the A_1 group, 7 out of 10 in the A_2 group and nearly 8 out of 10 in A_3 grew, implies very widespread hierarchical diffusion of manufacturing as measured by value added too. Growth did not bypass peri pheral areas altogether, but it was more spatially selective in the periphery of the national space-economy.

During the 1950-1960 decade the FEA cities did change, and there was a statistically-significant association with the extent of growth and decline in manufacturing value added in the counties associated with particular degree of change: growing counties had positive deviations, while they were negative for counties located in declinin or stagnating FEA's.

Dispersion

Relative accessibility of counties also was associated with the geographical distribution of change in manufacturing value added (Table 6.12). The deviation of observed frequencies from the expecte are as follows: first, both in 1950 and in 1960, counties in the high accessibility group (B_3) had a higher probability of experiencin growth in value added than counties in the other two categories of relative accessibility (B_2 and B_1); second, in 1950, the negative deviation was large only for the lowest accessibility group, B_1, but by 1960, substantial negative deviations occurred in both the B_1 and the B_2 categories. Such a change resulted from the expansion of the number of counties with high relative accessibility, resulting in a diminished number of counties with lower levels of relative accessibility, simultaneously reducing their posterior probability of growth

TABLE 6.12

Relative Accessibility (1950) and Change in Total Value-Added by Manufacturing 1952 - 1962.

Manufacturing Change	B_1		B_2		B_3		Total	P
Decline	409	(53.39)	358	(34.93)	274	(20.90)	1041	(33.56)
Growth	357	(46.61)	667	(65.07)	1037	(79.10)	2061	(66.44)
Total	766		1025		1311		3102	

	Expected Frequencies			Deviations From Observed Frequencies			Chi-Square Components		
	B_1	B_2	B_3	B_1	B_2	B_3	B_1	B_2	B_3
Decline	257	344	440	+152	+14	-166	89.90	.57	62.63
Growth	509	681	871	-152	-14	+166	45.39	.29	31.64

X^2 (2) = 230.43 Sig. < .001

Relative Accessiblity (1960) and Change in Total Value-Added by Manufacturing 1952 - 1962.

Manufacturing Change	B_1		B_2		B_3		Total	P
Decline	238	(50.21)	402	(46.26)	401	(22.80)	1041	(33.56)
Growth	236	(49.79)	476	(53.74)	1358	(77.20)	2061	(66.44)
Total	474		869		1759		3102	

	Expected Frequencies			Deviations From Observed Frequencies			Chi-Square Components		
	B_1	B_2	B_3	B_1	B_2	B_3	B_1	B_2	B_3
Decline	159	292	590	+79	+110	-189	39.25	41.44	60.54
Growth	315	577	1169	-79	-110	+189	19.81	20.97	30.56

Change in Relative Accessibility 1950-1960 and Change in Total Value-added by Manufacturing 1952 - 1962.

Manufacturing Change	b_1		b_2		Total	P
Decline	909	(32.86)	132	(39.41)	1041	(33.56)
Growth	1858	(67.14)	213	(60.59)	2061	(66.44)
Total	2767		335		3102	

	Expected Frequencies		Deviations From Observed Frequencies		Chi-Square Components	
	b_1	b_2	b_1	b_2	b_1	b_2
Decline	929	112	-20	+20	.43	3.57
Growth	1838	223	+20	-20	.22	1.79

X^2 (1) = 6.01 Sig. < .020

That is, the increased level of accessibility during the 1950's throughout the system resulted in dispersion of value added by manufacturing in all categories of relative accessibility, but the dispersion was particularly extensive in areas that had initially high relative accessibility or have achieved it by 1960.

Decentralization in Urban Fields

An almost perfect case of decentralization of manufacturing valued can be inferred from Table 6.13. In each commuting zone, more than one out of two counties experienced growth in manufacturing value added. The proportion of counties experiencing growth increase as the intensity of association with FEA city increased, conforming to an expected distance-decay function.

Hierarchical Diffusion and Dispersion

The hierarchy-national access interactions are familiar: Low relative accessibility exerts a negative influence on all hierarchical levels, high relative accessibility exerts positive influence on all hierarchical levels. However, only three cases of interaction are strong, A_1B_1, A_2B_3, and A_3B_3, reflecting a combination of two positive or two negative effects. These three deviations contribute more than 90 per cent to the overall chi-square value. The importanc of these three deviations suggest the advantage, in terms of growth probability, possessed by the economic core in 1950 and the disadvantage of the hard-core periphery (A_3B_3 vs. A_1B_1). At the same time all other deviations are relatively weak; indicating the pervasivenes of hierarchical diffusion and dispersion, and the system-maintaining effect of the interaction of these two processes. Such a maintenance effect fails at both ends of the spatial-economic system, however, implying significant changes. The strong negative effect extends from A_1B_1 in 1950 to A_1B_2 in 1960. Such posterior relative "deterioration" is attributable to the great expansion in the number of

TABLE 6.13

Relative Location in Urban Fields (1960) and Change in Total Value-Added by Manufacturing 1952 - 1962.

Manufacturing Change

	C_1	C_2	C_3	C_4	Total P
Decline	327 (49.77)	464 (36.16)	181 (27.98)	69 (13.40)	1041 (33.56)
Growth	330 (50.23)	819 (63.84)	466 (72.02)	446 (86.60)	2061 (66.44)
Total	657	1283	647	515	3102

	Expected Frequencies				Deviations From Observed Frequencies				Chi-Square Components			
	C_1	C_2	C_3	C_4	C_1	C_2	C_3	C_4	C_1	C_2	C_3	C_4
Decline	220	431	217	173	+107	+33	-36	-104	52.04	2.53	5.97	62.52
Growth	437	852	430	342	-107	-33	+36	+104	26.20	1.28	3.01	31.63

X^2 (3)= 185.18 Sig.$<$ .001

TABLE 6.14

Urban Hierarchy Interaction with Relative Accessibility (1960) and Change in Total Value-Added by Manufacturing 1952 - 1962.

Manufacturing Change

Interaction	Decline		Growth		Total
A_1B_1	168	(57.74)	123	(42.46)	291
A_1B_2	202	(59.77)	136	(40.23)	338
A_1B_3	99	(31.14	219	(68.84)	318
A_2B_1	49	(43.75)	63	(56.25)	112
A_2B_2	92	(33.18)	149	(61.82)	241
A_2B_3	150	(23.44)	490	(76.56)	640
A_3B_1	21	(29.58)	50	(70.42)	71
A_3B_2	108	(37.25)	182	(62.75)	290
A_3B_3	152	(18.98)	649	(81.02)	801
Total	1041	(33.56)	2061	(66.44)	3102

	Expected Frequencies		Deviations From Observed Frequencies		Chi-Square Components	
	Decline	Growth	Decline	Growth	Decline	Growth
A_1B_1	98	193	+70	-70	50.00	25.39
A_1B_2	113	225	+89	-89	70.10	35.20
A_1B_3	107	211	- 8	+ 8	.60	.30
A_2B_1	38	74	+11	-11	3.18	1.64
A_2B_2	81	160	+11	-11	1.49	.76
A_2B_3	215	425	-65	+65	19.65	9.94
A_3B_1	24	47	- 3	+ 3	.38	.19
A_3B_2	97	193	+11	-11	1.25	.63
A_3B_3	268	533	-116	+116	50.21	25.25

x^2 (8)= 296.16 Sig.< .001

Interaction of Change in F.E.A City Size with Change in Relative Accessibility (1950 - 1960) and Change in Total Value-Added by Manufacturing 1952 - 1962.

Manufacturing Change

Interaction	Decline		Growth		Total
a_1b_1	251	(51.12)	240	(48.88)	491
a_1b_2	71	(61.74)	44	(38.26)	115
a_2b_1	113	(19.86)	456	(80.14)	569
a_2b_2	2	(8.70)	21	(91.30)	23
a_3b_1	236	(28.96)	579	(71.04)	815
a_3b_2	18	(32.73)	37	(67.27)	55
a_4b_1	309	(34.65)	583	(65.35)	892
a_4b_2	41	(28.88)	101	(71.12)	142
Total	1041	(33.56)	1156	(66.44)	3102

	Expected Frequencies		Deviations From Observed Frequencies		Chi-Square Components	
	Decline	Growth	Decline	Growth	Decline	Growth
a_1b_1	165	326	+86	-86	44.82	22.69
a_1b_2	39	76	+32	-32	26.26	13.47
a_2b_1	191	378	-78	+78	31.85	16.10
a_2b_2	8	15	- 6	+ 6	4.50	2.40
a_3b_1	274	541	-38	+38	5.27	2.67
a_3b_2	18	37	--	--	----	----
a_4b_1	299	593	+10	-10	.33	.17
a_4b_2	47	95	- 6	+ 6	.77	.38

x^2 (7)= 171.68 Sig.< .001

TABLE 6.14
(cont.)

Urban Hierarchy Interaction with Relative Accessibility (1950) and the Change in Total Value-Added by Manufacturing 1952 - 1962.

Interaction	Manufacturing Change: Decline		Manufacturing Change: Growth		
A_1B_1	302	(61.51)	189	(38.49)	491
A_1B_2	147	(41.88)	204	(58.12)	351
A_1B_3	62	(27.43)	164	(73.57)	226
A_2B_1	75	(44.38)	94	(55.62)	169
A_2B_2	119	(32.34)	249	(67.66)	368
A_2B_3	109	(22.61)	373	(77.39)	482
A_3B_1	32	(30.19)	74	(69.81)	106
A_3B_2	92	(30.06)	214	(69.94)	306
A_3B_3	103	(17.08)	500	(82.92)	603
Total	1041	(33.56)	2061	(66.44)	3102

	Expected Frequencies		Deviations From Observed Frequencies		Chi-Square Components	
	Decline	Growth	Decline	Growth	Decline	Growth
A_1B_1	165	326	+137	-137	113.75	57.57
A_1B_2	118	233	+29	-29	7.12	3.61
A_1B_3	76	150	-14	+14	2.58	1.31
A_2B_1	57	112	+18	-18	5.68	2.89
A_2B_2	123	245	- 4	+ 4	.13	.06
A_2B_3	162	320	-53	+53	17.34	8.78
A_3B_1	36	70	- 4	+ 4	.44	.23
A_3B_2	103	203	-11	+11	1.17	.60
A_3B_3	201	402	-98	+98	47.78	23.89

X^2 (8)= 294.93 Sig.<.001

counties with high relative accessibility, and the relatively increased peripherality of the A_1B_2 spatial group.

When we examine the association between the interaction of hierarichical change with change in relative accessibility and the change in value added, it is clear that the acceptance of the association is due largely to three groups of counties, a_1b_1, a_1b_2, and a_2b_1, which account for more than 90 per cent of the overall chi-square value. The direction of the deviation from expected frequencies is negative for a_1b groups. Stagnant FEAs experienced decline in manufacturing value added. Such a result holds regardless of the change in relative accessibility. Indeed, there are grounds to believe that a majority of the a_1b counties are in the peripheral areas of the nation, with no major urban center in the economic region.

Hierarchical Diffusion and Decentralization

The simultaneous effect of interactions between hierarchical status and relative location in the urban field is revealed in Table 6.15. Looking at the first part of the table, it seems hard to find any systematic variation or progression of frequencies. The only exception is the low frequency of growth in A_1C_1 and A_1C_2 groups and an apparent increase in such frequency as one moves to A_1C_3 and A_1C_4. The second part of the table confirms the high negative deviation from expected values of the A_1C_1 and A_1C_2 groups. In fact, these two groups, including less than 30 per cent of all counties in the U.S., contribute 50 per cent of the overall chi-square value. Thus, although growth in manufacturing value added occurred throughout the system, including the peripheral groups of counties A_1C_1 and A_1C_2, location at the periphery of urban fields of small urban centers implied a smaller probability of growth and geographically more selective industrialization. However, that situation just describes changes in counties closer to the small urban centers, leading to

TABLE 6.15

Urban Hierarchy (1950) Interaction with Relative Location in Urban Field (1960) and Change in Total Value-Added by Manufacturing 1952-1962.

Interaction	Manufacturing Change: Decline		Manufacturing Change: Growth		Total
A_1C_1	323	(53.39)	282	(46.61)	605
A_1C_2	126	(55.75)	100	(44.25)	226
A_1C_3	47	(36.43)	82	(63.57)	129
A_1C_4	15	(13.89)	93	(86.11)	108
A_2C_1	2	(11.76)	15	(88.24)	17
A_2C_2	197	(38.78)	311	(61.22)	508
A_2C_3	68	(24.64)	208	(75.36)	276
A_2C_4	36	(16.51)	182	(83.49)	218
A_3C_1	2	(5.71)	33	(94.29)	35
A_3C_2	141	(25.68)	408	(74.32)	549
A_3C_3	66	(27.27)	176	(72.73)	242
A_3C_4	18	(9.52)	171	(91.48)	189
Total	1041	(33.56)	2061	(66.44)	3102

	Expected Frequencies		Deviations From Observed Frequencies		Chi-Square Components	
	Decline	Growth	Decline	Growth	Decline	Growth
A_1C_1	203	402	+120	-120	70.94	35.82
A_1C_2	76	150	+50	-50	32.89	16.67
A_1C_3	43	86	+ 4	- 4	.37	.19
A_1C_4	36	72	-21	+21	12.25	6.12
A_2C_1	6	11	- 4	+ 4	2.67	1.45
A_2C_2	170	338	+27	-27	4.29	2.16
A_2C_3	93	183	-25	+25	6.72	3.42
A_2C_4	73	145	-37	+37	18.75	9.44
A_3C_1	12	23	-10	+10	8.33	4.35
A_3C_2	184	365	-43	+43	10.05	5.07
A_3C_3	81	161	-15	+15	2.78	1.40
A_3C_4	64	125	-46	+46	33.06	16.93

X^2 (9)= 306.12 Sig.<.001

TABLE 6.15
(cont.)

Urban Hierarchy Interaction with Relative Location in Urban Field (1960) and Change in Total-Added by Manufacturing 1952 - 1962.

Interaction	Manufacturing Change: Decline		Growth		Total
A_1C_1	323	(53.57)	280	(46.43)	603
A_1C_2	95	(56.22)	74	(43.78)	169
A_1C_3	39	(39.80)	59	(60.20)	98
A_1C_4	12	(15.59)	65	(84.41)	77
A_2C_1	1	(5.56)	17	(94.44)	18
A_2C_2	193	(39.55)	295	(60.45)	488
A_2C_3	66	(32.52)	203	(67.48)	269
A_2C_4	31	(14.22)	187	(85.78)	218
A_3C_1	3	(8.34)	33	(91.66)	36
A_3C_2	176	(28.12)	450	(71.88)	626
A_3C_3	76	(27.15)	204	(72.85)	280
A_3C_4	26	(11.82)	194	(88.18)	220
Total	1041	(33.56)	2061	(66.44)	3102

	Expected Frequencies		Deviations From Observed Frequencies		Chi-Square Components	
	Decline	Growth	Decline	Growth	Decline	Growth
A_1C_1	202	401	+121	-121	72.48	36.51
A_1C_2	57	112	+38	-38	25.33	12.89
A_1C_3	33	65	+ 6	- 6	1.09	.55
A_1C_4	26	51	-14	+14	7.54	3.84
A_2C_1	6	12	- 5	+ 5	4.17	2.08
A_2C_2	164	324	+29	-29	5.13	2.60
A_2C_3	90	179	-24	+24	6.40	3.22
A_2C_4	73	145	-42	+42	24.16	12.17
A_3C_1	12	24	- 9	+ 9	6.75	3.38
A_3C_2	210	416	-34	+34	5.50	2.78
A_3C_3	94	192	-18	+18	3.45	1.69
A_3C_4	74	146	-48	+48	31.14	15.78

X^2 (9)= 290.63 Sig.<.001

TABLE 6.15
(cont.)

Interaction of Change in F.E.A. City Size (1950 - 1960) with Relative Location in Urban Field (1960) and Change in Total Value-Added by Manufacturing 1952 - 1962.

Interaction	Manufacturing Change: Decline		Manufacturing Change: Growth		Total
$a_1+a_2C_1+C_2$	389	(42.33)	530	(57.67)	919
$a_1+a_2C_3$	34	(22.82)	115	(77.18)	149
$a_1+a_2C_4$	14	(10.77)	116	(89.23)	130
$a_3C_1+C_2$	160	(37.47)	267	(62.53)	427
a_3C_3	68	(27.31)	181	(72.69)	249
a_3C_4	26	(13.41)	168	(86.59)	194
$a_4C_1+C_2$	242	(40.74)	352	(59.26)	594
a_4C_3	79	(31.73)	170	(68.27)	249
a_4C_4	29	(15.19)	162	(84.81)	191
Total	1041	(33.56)	2061	(66.44)	3102

	Expected Frequencies		Deviations From Observed Frequencies		Chi-Square Components	
	Decline	Growth	Decline	Growth	Decline	Growth
$a_1+a_2C_1+C_2$	308	611	+81	-81	21.30	10.74
$a_1+a_2C_3$	50	99	-16	+16	5.12	2.59
$a_1+a_2C_4$	44	86	-30	+30	20.45	10.46
$a_3C_1+C_4$	143	284	+17	-17	2.02	1.02
a_3C_3	84	165	-16	+16	3.05	1.55
a_3C_4	65	129	-39	+39	23.40	11.79
$a_4C_1+C_2$	199	395	+43	-43	9.29	4.68
a_4C_3	83	166	- 4	+ 4	.19	.10
a_4C_4	65	126	-36	+36	19.94	10.29

X^2 (8)= 157.98 Sig.< .001

the conclusion that relative location in urban fields is more crucial in lower-status FEA's than it is in higher status areas. This is not to say that relative location in the urban field is not important for the A_2 and A_3 groups, but that it seems to be relatively less so. Indeed, two highly positive deviations should be noted: A_2C_4 and A_3C_4. Further, the systematically high positive deviations of the AC_4 groups should indicate the positive effect of being located close to an urban center and of the intensive interaction between county and central city, regardless of the absolute economic size of such a center.

These relationships have a constancy at the later period of time 1960. Examining the change in value added by manufacturing from *a posteriori* levels of location reveals no major changes in the AC effects on frequencies.

Looking at the aC interaction effect in the third part of Table 6.15 reveals the strong effect of C. First, note the negative deviation of observed frequencies of the aC_1+C_2 groups. The negative deviation is systematic; only the strength of the deviations varies. It is stronger in areas where the FEA city stagnated or declined, and is weaker where growth of FEA city was average. Secondly, the relatively higher importance of C than of a in the interaction aC is again demonstrated by the systematically strong positive deviation of the three groups aC_4. Finally, a weak effect is generally exerted on the C_3, the transition zone between the urban fringe and the rural areas. In conclusion, one suspects that the relative location of counties in urban fields affects not only the frequency of growth in value added but probably its intensity also.

Dispersion and Decentralization

The distribution of growth and decline of value added for both the 1950 and 1960 categories of relative accessibility with respect to relative location within urban fields is relatively constant (Table 6.16). Examination of the chi-square deviations reveals the following: negative deviations in $B_1C_{1,2,3}$ and in $B_2C_{1,2,3}$ and positive deviations for all other groups. The difference in the relative significance of the deviations suggests that C_1, C_2, and C_3 exert a negative influence, in decreasing order, for any given B level. C_4 on the other hand, exerts a positive influence. Similarly, the effect of B_1 and B_2 on any given C level is negative while the effect of B_3 is positive. Once again it seems that the controlling locational component most crucial in the interactions AC and BC is the relative location in the urban field, component C, rather than A or B. The differences between 1950 and 1960 are simply that, as the periphery has shrunk, the growth proportions have shifted:

	1950 No. of counties	Growth Proportion	1960 No. of counties	Growth proportion
B_1C_1	335	36.12	234	40.59
B_1C_2	250	40.80	141	43.93
B_1C_3	84	57.14	47	70.21

Thus, as the areas of poor relative accessibility diminished, those that remained with poor relative accessibility nationally and locally improved in terms of frequency of occurrence of growth in value added by manufacturing. The reverse is true for the $B_2C_{1,2,3}$ groups. What is the meaning of such a shift? It is possible to speculate and say that improved accessibility resulted in a relatively more equitable distribution of the proportion of counties that experienced growth or decline in value added by manufacturing in areas with low and medium level of relative accessibility (B_1 or B_2) than would have been the case if there was no or little change in relative accessibility during the 1950's. That is, it seems that continued population growth and improved communications, both components of relative accessibility of counties, could be the crucial factor in reducing

TABLE 6.16

Relative Accessibility (1950) Interaction with Relative Location in Urban Field (1960) and Change in Total Value-Added by Manufacturing 1952 - 1962.

Interaction	Manufacturing Change: Decline		Growth		Total
B_1C_1	214	(63.88)	121	(36.12)	335
B_1C_2	148	(59.20)	102	(40.80)	250
B_1C_3	36	(42.86)	48	(57.14)	84
B_1C_4	11	(11.34)	86	(88.64)	97
B_2C_1	73	(40.78)	106	(59.22)	179
B_2C_2	176	(36.59)	305	(63.41)	481
B_2C_3	82	(36.94)	140	(63.06)	222
B_2C_4	27	(18.88)	116	(81.12)	143
B_3C_1	40	(27.97)	103	(72.03)	143
B_3C_2	140	(25.36)	412	(74.64)	552
B_3C_3	63	(18.48)	278	(81.52)	341
B_3C_4	31	(11.27)	244	(88.73)	275
Total	1041	(33.56)	2061	(66.44)	3102

	Expected Frequencies		Deviations From Observed Frequencies		Chi-Square Components	
	Decline	Growth	Decline	Growth	Decline	Growth
B_1C_1	112	223	+102	-102	92.89	46.65
B_1C_2	84	166	+64	-64	48.76	24.67
B_1C_3	28	56	+ 8	- 8	2.29	1.14
B_1C_4	33	64	-22	+22	14.67	7.56
B_2C_1	60	119	+13	-13	2.81	1.42
B_2C_2	161	320	+15	-15	1.40	.70
B_2C_3	74	148	+ 8	- 8	.86	.43
B_2C_4	48	95	-21	+21	9.19	4.64
B_3C_1	48	95	- 8	+ 8	1.33	.67
B_3C_2	185	367	-45	+45	10.94	5.52
B_3C_3	114	227	-51	+51	22.81	1 .46
B_3C_4	94	181	-63	+63	42.22	21.93

X^2 (11)= 366.96 Sig.<.001

TABLE 6.16
(cont.)

Relative Accessibility Interaction with Relative Location in Urban Field (1960) and Change in Total Value-Added by Manufacturing.

Interaction	Manufacturing Change: Decline		Manufacturing Change: Growth		Total
B_1C_1	139	(59.41)	95	(40.59)	234
B_1C_2	79	(56.03)	62	(43.93)	141
B_1C_3	114	(29.79)	33	(70.21)	47
B_1C_4	6	(11.54)	46	(88.46)	52
B_2C_1	116	(58.59)	82	(41.41)	198
B_2C_2	192	(47.65)	211	(52.35)	403
B_2C_3	72	(48.65)	76	(51.35)	148
B_2C_4	22	(18.34)	98	(81.66)	120
B_3C_1	72	(32.00)	153	(68.00)	225
B_3C_2	193	(26.12)	546	(73.88)	739
B_3C_3	95	(21.02)	357	(78.98)	452
B_3C_4	41	(11.96)	302	(88.04)	343

	Expected Frequencies		Deviations From Observed Frequencies		Chi-Square Components	
	Decline	Growth	Decline	Growth	Decline	Growth
B_1C_1	78	154	+61	-61	47.70	24.16
B_1C_2	47	94	+32	-32	21.79	10.89
B_1C_3	16	31	- 2	+ 2	.25	12.91
B_1C_4	17	35	-11	+11	7.12	3.46
B_2C_1	66	132	+50	-50	37.88	18.94
B_2C_2	135	268	+57	-57	24.07	12.12
B_2C_3	50	98	+22	-22	9.68	4.94
B_2C_4	40	80	-18	+18	8.10	4.05
B_3C_1	76	149	- 4	+ 4	.21	.11
B_3C_2	248	491	-55	+55	12.20	6.16
B_3C_3	152	300	-57	+57	21.38	10.83
B_3C_4	116	227	-75	+75	48.49	24.78

X^2 (11)= 372.22 Sig.< .001

TABLE 6.16
(cont.)

Interaction of Change in Relative Accessibility (1950 - 1960) with Relative Location in Urban Field and Change in Total Value-Added by Manufacturing 1952 - 1962.

Interaction	Manufacturing Change				
	Decline		Growth		Total
b_1C_1	255	(47.67)	280	(52.33)	535
b_1C_2	425	(35.48)	773	(64.52)	1198
b_1C_3	165	(27.37)	438	(72.63)	603
b_1C_4	64	(14.85)	367	(85.15)	431
b_2C_1	72	(59.02)	50	(40.98)	122
b_2C_2	39	(45.89)	46	(54.11)	85
b_2C_3	16	(36.37)	28	(63.63)	44
b_2C_4	5	(5.96)	79	(94.04)	84
Total	1041	(33.56)	2061	(66.44)	3102

	Expected Frequencies		Deviations From Observed Frequencies		Chi-Square Components	
	Decline	Growth	Decline	Growth	Decline	Growth
b_1C_1	180	355	+75	-75	31.25	15.84
b_1C_2	402	796	+23	-23	1.32	.66
b_1C_3	202	401	-37	+37	6.78	3.41
b_1C_4	145	286	-81	+81	45.25	22.94
b_2C_1	41	81	+31	-31	23.44	11.86
b_2C_2	28	57	+11	-11	4.32	2.12
b_2C_3	15	29	+ 1	- 1	.04	.03
b_2C_4	28	56	-23	+23	18.89	9.45

X^2 (7)= 197.60 Sig.< .001

spatial inequality - at least in the geographical extent of growth and decline.

Finally, the bC interaction effect shown in Table 6.16 indicates the relative insignificance of b as compared to the relative location of counties in their urban field. For any given level of relative accessibility change there are two important C effects. Location at the periphery of an urban field (C_1) resulted in a negative impact on the proportion of counties experiencing growth. Location at the core of the urban field, the area of intense interaction between county and FEA city, resulted in a positive influence.

Relative Location and Change Differentials

The thing to note in the analysis of variance of the rate of change in value-added by manufacturing 1952-1962 is the fact that the relative location effects, both main and interaction effects, are not very powerful. That is, although the effects are statistically significant, the F-values are low, particularly when they are compared to the values in other variables in this study.

Examining Table 6.17 first, the most interesting result is the lack of significance of relative accessibility. In other words, contrary to expectation, initial levels of relative accessibility are not statistically associated with the degree of change in value added by manufacturing. The conclusion that B is insignificant as a main effect leaves us with two significant main effects: the hierarchical effect and the decentralization effect. Of the two, hierarchical diffusion is the most important process, and the AB and AC interaction effects remain statistically significant.

The direction of influence of each subeffect is given in Table 6.17. Thus, association with a small urban center in 1950 implied a greater likelihood of a low rate of change in value added. Similarly, a county located in the urban field of a small center that

TABLE 6.17

Analysis of Variance--1950

Effect	Mean Square	Degrees of Freedom	F. Ratio	Significance
A	93.11	2 ,3066	5.67	<.0035
B	12.06	2 ,3066	.73	<.4817
C	58.93	3 ,3066	3.59	<.0131
AB	90.65	4 ,3066	5.52	<.0002
AC	34.76	6 ,3066	2.12	<.0479
BC	58.06	6 ,3066	3.54	<.0018
ABC	14.37	12 ,3066	.88	<.5729

Dependent variable variance is 16.42 with 3066 degrees of freedom

ANOVA: Each Effect Alone

Effect	Mean Square	Degrees of Freedom	F. Ratio	Significance
A	93.11	2 ,3066	5.67	<.0035
B	40.58	2 ,3066	2.47	<.0842
C	87.37	3 ,3066	5.32	<.0012
AB	48.45	4 ,3066	2.95	<.0190
AC	43.07	6 ,3066	2.62	<.0153
BC	44.38	6 ,3066	2.70	<.0127
ABC	22.72	12 ,3066	1.38	<.1650

ANOVA: Significant Effects Only

	Effect	Mean Square	Degrees of Freedom	F. Ratio	Significance
1.	A	93.11	2 ,3066	5.67	<.0035
	C	62.56	3 ,3066	3.81	<.0096
	AC	36.26	6 ,3066	2.21	<.0391
	AB	78.12	4 ,3066	4.76	<.0008
2.	C	87.37	3 ,3066	5.32	<.0012
	A	55.89	2 ,3066	3.40	<.0332
	AC	36.26	6 ,3066	2.09	<.0391
	AB	78.12	4 ,3066	4.76	<.0008
3.	AB	48.45	4 ,3066	2.95	<.0190
	C	84.03	3 ,3066	5.12	<.0016
	A	124.16	2 ,3066	7.76	<.0006
	AC	34.96	6 ,3066	2.13	<.0466

Least Square Estimates of Effects

General Mean	+2.63
A_1	- .74
A_2	- .35
C_1	+ .18
C_2	- .86
C_3	- .46
A_1C_1	-1.06
A_1C_2	+2.10
A_1C_3	- .68
A_2C_1	+ .58
A_2C_2	+ .20
A_2C_3	+ .76
A_1B_1	-2.05
A_1B_2	-1.31
A_2B_1	-1.63
A_2B_2	- .75

possessed low potential was likely to experience less growth in value added by manufacturing than a county associated with the same hierarchical level but with higher population potential. Inferences can be drawn in the same way for each subeffect.

In terms of processes one cannot but conclude that hierarchical diffusion was the most important process determining change in value added by manufacturing. The dispersion process was less significant, while decentralization occurred primarily in terms of spread effects through the urban hierarchy. It seems that high value added industry decentralized to smaller urban centers and to nearby counties, and to areas that had achieved threshold conditions for the support of manufacturing plants. Yet, the dominance of the core manufacturing areas in terms of change in value added by manufacturing also remained unquestioned.

What was the effect of the change in relative location during the 1950-1960 period on the change in value added in manufacturing during the 1952-1962 period? Table 6.19 provides interesting insights. Four factors in a sequence are associated with the change in value added: the change in the population of the urban centers in the urban hierarchy (a); the change in the county population potential (b); the 1960 constant of relative location within the urban field (c); and the interaction factor of change in FEA city population and in the county's population potential (ab). In particular, the change in population potential seems to be the most important main factor.

A different perspective is added by examining the interaction between initial conditions, 1950 relative location, and the changes in such conditions, 1950-1960, however. Because of the detailed breakdown, three factors emerge as statistically significant. The interaction between FEA city population size and the change in population, the interaction between population potential of counties in 1950 and the change in it during the 1950-1960 period and relative

TABLE 6.18

Analysis of Variance (1960)

Effect	Mean Square	Degrees of Freedom	F. Ratio	Significance
A	108.19	2 ,3066	6.62	<.0014
B	56.72	2 ,3066	2.97	<.2782
C	50.77	3 ,3066	3.10	<.0253
AB	60.76	4 ,3066	3.72	<.0051
AC	28.85	6 ,3066	1.76	<.1017
BC	52.51	6 ,3066	3.21	<.0038
ABC	38.72	12 ,3066	2.37	<.0050

ANOVA: Each Effect Alone

Effect	Mean Square	Degrees of Freedom	F. Ratio	Significance
A	108.19	2 ,3066	6.62	<.0014
B	87.54	2 ,3066	5.35	<.0048
C	87.37	3 ,3066	5.34	<.0012
AB	29.79	4 ,3066	1.82	<.1210
AC	32.24	6 ,3066	1.97	<.0657
BC	41.76	6 ,3066	2.55	<.0180
ABC	25.47	12 ,3066	1.56	<.0964

ANOVA: Significant Effects Only

Effect	Mean Square	Degrees of Freedom	F. Ratio	Significance
C	87.37	3 ,3066	5.34	<.0012
BC	61.84	6 ,3066	3.78	<.0010
A	65.07	2 ,3066	3.98	<.0187
B	39.38	2 ,3066	2.41	<.0895
A	108.19	2 ,3066	6.62	<.0014
C	57.02	3 ,3066	3.49	<.0150
AB	54.88	4 ,3066	3.36	<.0095

Least Square Estimates of Effects

General Mean	+2.34
C_1	- .76
C_2	-1.04
C_3	- .43
B_1C_1	-1.10
B_1C_2	-2.17
B_1C_3	-1.07
B_2C_1	-1.58
B_2C_2	+ .44
B_2C_3	-1.97
A_1	- .64
A_2	- .20
B_1	+ .24
B_2	+ .32

TABLE 6.19

Analysis of Variance (1950-1960)

Effect	Mean Square	Degrees of Freedom	F. Ratio	Significance
a	65.60	3 ,3074	3.98	<.0076
b	222.73	1 ,3074	13.52	<.0003
C	54.58	3 ,3074	3.31	<.0191
ab	99.76	3 ,3074	6.06	<.0005
aC	10.10	9 ,3074	.61	<.7875
bC	40.69	3 ,3074	2.47	<.0597
abC	N.A.			

ANOVA: Each Effect Alone

Effect	Mean Square	Degrees of Freedom	F. Ratio	Significance
a	65.59	3 ,3074	3.98	<.0076
b	163.52	1 ,3074	9.93	<.0017
C	87.37	3 ,3074	5.30	<.0012
ab	12.83	3 ,3074	.78	<.5086
aC	49.30	9 ,3074	2.99	<.0015
bC	15.83	3 ,3074	.96	<.4131
abC	11.09	9 ,3074	.67	<.7340

Other subtables N.A.

location in urban field as represented by the spatial constant C, all are significant. This is confirmed throughout Table 6.20. Changing the order of these three factors in the three combinations available does not change their statistical significance. One can choose any of the three combinations available:

1. $Aa \rightarrow Bb \rightarrow C$
2. $Aa \rightarrow C \rightarrow Bb$
3. $C \rightarrow Bb \rightarrow Aa$

and still the statistical significance holds.

The interesting analysis is that of the direction of influence of the processes involved. Thus, counties associated with small FEA cities in 1950 that experienced rapid population increase were likely to experience well above average rates of change in value added in manufacturing. This was not so with respect to A_1a_3, a result that could imply that during the 1950's and early sixties rapid urbanization led to rapid increase in value added by manufacturing. It also points out the concentration of value added growth in the most rapidly urbanizing rural areas. Such a conclusion is further confirmed by the direction of operation of the C factors, indicating that above-average change in value added by manufacturing could be expected in those counties that by 1960 achieved high commuting levels to the FEA cities regardless of their hierarchical level (C_1, C_2, and C_3 all have negative effects and only C_4 a positive effect). The A_1a direction of effect reverses itself, A_2a_1, A_2a_2 and A_2a_3 exerting an indecisive positive effect while A_2a_4 exerts a negative effect. That is, counties associated in 1950 with medium size FEA cities and experiencing rapid growth in population were unlikely to achieve above-average change in value added by manufacturing during 1952-1962. Finally, the Bb terms involving initial relative accessibility and its change show that counties that had only low initial population potential and only average changes had a highly negative effect on

TABLE 6.20

Analysis of Variance-1950 and 1950-1960

Effect	Mean Square	Degrees of Freedom	F. Ratio	Significance
Aa	56.05	6 ,2959	3.41	<.0024
Bb	70.46	2 ,2959	4.28	<.0139
C	45.90	3 ,2959	2.79	<.0388
AaBb	25.99	12 ,2959	1.58	<.0898
AaC	18.42	18 ,2959	1.12	<.3237
BbC	24.23	6 ,2959	1.47	<.1828

ANOVA: Each Effect Alone

Effect	Mean Square	Degrees of Freedom	F. Ratio	Significance
Aa	56.05	6 ,2959	3.41	<.0024
Bb	108.44	2 ,2959	6.59	<.0014
C	87.37	3 ,2959	5.31	<.0012
AaBb	26.90	12 ,2959	1.64	<.0749
AaC	23.48	18 ,2959	1.43	<.1074
BbC	12.38	6 ,2959	.75	<.6077

ANOVA: Significant Effects Only

	Effect	Mean Square	Degrees of Freedom	F. Ratio	Significance
1.	Aa	56.06	6 ,2959	3.41	<.0024
	C	52.73	3 ,2959	3.21	<.0221
	Bb	60.22	2 ,2959	3.66	<.0258
2.	Aa	56.05	6 ,2959	3.41	<.0024
	Bb	70.46	2 ,2959	4.28	<.0139
	C	45.90	3 ,2959	2.79	<.0388
3.	C	87.37	3 ,2959	5.31	<.0012
	Bb	73.27	2 ,2959	4.46	<.0117
	Aa	34.39	6 ,2959	2.09	<.0510

Least Square Estimates of Effects

General Mean	+2.31
A_1a_1	- .44
A_1a_2	+ .44
A_1a_3	-1.03
A_2a_1	+1.55
A_2a_2	+ .17
A_2a_3	+ .01
C_1	- .37
C_2	- .61
C_3	- .36
B_1b_1	-1.13
B_2b_1	+ .35

the change in value added by manufacturing. If however, there was <u>rapid</u> change in population potential the opposite was likely: rapid positive change in value added by manufacturing.

CHAPTER 7

THE CONDITIONAL NATURE OF MANUFACTURING GROWTH

The problem remaining is one of integration and synthesis, of cutting through the statistical tedium of Chapters 2-6 and developing an overview of the geographical change in U.S. manufacturing during the 1950's, as well as an understanding of the spatial processes involved. First, recall from the discussion in Chapter 2 that one feature of the decade was the rapid improvement in the relative location of most counties in the country. By 1960 there were many fewer low-accessibility counties and low-status urban centers than there were in 1950. To the extent that the probability of growth varied positively with relative location, then, counties changing their relative location -- filtering up -- improved their growth prospects, and simultaneously, the area of the high-accessibility national heartland expanded at the expense of the periphery, high-status regions grew at the expense of low-status regions, and zones of superior local access grew at the expense of outlying areas of inferior access. The relative location changes we have characterized as "filtering up" have also been termed "spread" or "backwash" effects, the spillover of growth from heartland centers into adjacent areas of what was formerly the periphery. Such effects involve dispersion, diffusion and decentralization in two ways:

(a) In the small but consistently positive growth probability of the less-accessible, low-status, peripheral areas.

(b) In the shifts of relative location of counties within the nation's economic space, implying a move of a county from one set of growth probabilities to another.

But there is a third way in which the three redistributive processes can work as well:

(c) By changes in growth probabilities in favor of the smaller peripheral regions.

It is important that these cases be clearly differentiated in any geographical analysis of manufacturing change, and we will attempt to do so in what follows.

We begin by discussing a table which summarizes the principal conclusions about variations in the growth rates of each of the seven manufacturing variables (Table 7.1).

For all categories of relative location, including both the main and the interaction effects, one can see in this table that positive growth probabilities existed even for the most extreme periphery of the country. Thus, with respect to change in the total number of manufacturing establishments, no locational category had a growth proportion of less than 50 percent. That is, no matter in what area of the U.S. one looked for industrial change, the probability of growth during the 1950's was greater than one in two.

But at the same time, growth probabilities were lower in peripheral areas than in the core areas of the country; while diffusion, decentralization and dispersion were taking place, the heartland continued to accumulate manufacturing at a rapid rate.

One variable showed a reverse pattern of growth probabilities however: change in manufacturing employment by place of residence. In this case, the probability of growth in the peripheral areas was much larger than the growth probability in heartland core areas, indicating an accelerated industrialization of the periphery at least

TABLE 7.1

Proportions of Counties Growing: 1950, 1960, and 1950 - 1960

1950 Growth Concentration

Manufacturing Change Variable	All Establish-ments	Large Establish-ments	Medium Size Establish-ments	Employment Place of Residence	Employment Place of Work	Payroll	Value-Added	Number of Counties
A								
A_1	62.83	13.67	24.34	70.13	52.25	54.31	52.15	1068
A_2	74.19	25.22	38.27	58.98	61.83	74.09	70.26	1019
A_3	79.41	27.78	44.43	58.42	63.06	80.20	77.64	1015
B								
B_1	61.23	11.88	23.11	65.53	49.74	48.96	46.61	766
B_2	70.34	19.12	33.95	70.63	59.51	68.29	65.07	1025
B_3	79.56	30.36	43.94	54.69	63.84	81.92	79.10	1311
C								
C_1	61.19	14.92	21.46	71.53	59.68	53.06	50.23	657
C_2	70.46	18.24	32.50	68.51	59.39	67.42	63.84	1283
C_3	74.81	24.27	37.56	60.74	60.12	76.20	72.02	647
C_4	86.02	38.06	58.25	39.03	66.80	87.18	86.60	515
AB								
A_1B_1	53.16	6.92	16.09	68.84	42.36	40.53	38.49	491
A_1B_2	66.10	13.96	29.92	78.06	54.42	60.11	58.12	351
A_1B_3	78.76	27.88	33.63	60.72	70.05	75.22	73.57	266
A_2B_1	69.82	20.12	31.36	56.80	60.95	59.17	55.62	169
A_2B_2	70.65	22.01	33.15	67.66	62.50	72.28	67.66	368
A_2B_3	78.42	29.46	44.61	53.11	61.62	80.70	77.39	418
A_3B_1	84.91	21.70	42.45	65.15	66.04	71.70	69.81	106
A_3B_2	74.84	21.57	39.54	65.68	61.76	72.88	69.94	306
A_3B_3	80.76	32.01	47.26	53.77	63.18	85.41	82.92	603
AC								
A_1C_1	59.34	11.44	19.34	72.89	49.19	48.60	46.61	605
A_1C_2	59.74	7.52	24.78	74.33	48.23	47.79	44.25	226
A_1C_3	67.44	20.93	23.26	67.44	58.92	65.12	63.57	129
A_1C_4	83.33	28.70	52.78	49.07	70.37	87.04	86.11	108
A_2C_1	82.35	58.82	58.82	52.94	70.59	88.24	88.24	17
A_2C_2	69.69	17.52	31.30	67.32	60.43	66.14	61.22	508
A_2C_3	74.91	25.34	39.13	57.97	60.87	80.80	75.36	276
A_2C_4	84.40	40.37	51.84	41.28	65.60	83.03	83.49	218
A_3C_1	82.86	48.57	40.00	57.14	68.57	94.29	94.29	35
A_3C_2	75.59	23.32	36.80	67.21	63.02	76.68	74.32	549
A_3C_3	79.75	24.79	43.49	60.33	59.92	76.86	72.73	242
A_3C_4	89.42	40.74	68.78	30.08	66.14	92.06	91.48	189
BC								
B_1C_1	50.75	5.08	12.54	72.23	38.81	37.61	36.12	335
B_1C_2	62.40	5.60	21.60	70.80	47.60	44.40	40.80	250
B_1C_3	67.86	20.24	29.76	61.90	61.90	61.90	57.14	84
B_1C_4	88.66	44.33	57.73	31.95	82.48	88.66	88.64	97
B_2C_1	65.92	16.76	27.38	79.33	59.22	61.45	59.22	179
B_2C_2	68.19	14.76	32.43	76.09	60.08	66.94	63.41	481
B_2C_3	68.92	15.77	30.63	68.01	51.35	68.02	63.06	222
B_2C_4	85.32	41.96	52.45	45.45	70.63	81.82	81.12	143
B_3C_1	79.72	35.66	34.97	60.14	67.83	74.13	72.03	143
B_3C_2	76.19	26.99	37.50	60.87	64.13	78.26	74.64	552
B_3C_3	80.35	30.79	43.99	56.01	65.40	85.04	81.52	341
B_3C_4	85.46	33.82	61.46	37.81	59.27	89.45	88.73	275

1960 Growth Concentration

Manufacturing Change Variable	All Establish-ments	Large Establish-ments	Medium Size Establish-ments	Employment Place of Residence	Employment Place of Work	Payroll	Value-Added	Number of Counties
A								
A_1	61.56	12.04	22.81	71.91	50.79	52.80	50.47	947
A_2	75.83	25.18	38.07	59.31	62.74	74.22	70.69	993
A_3	77.19	27.62	43.63	57.91	62.31	78.49	71.81	1162
B								
B_1	64.56	12.03	24.05	61.18	50.63	52.53	49.79	474
B_2	65.25	13.58	28.77	72.61	53.97	56.27	53.74	869
B_3	77.31	28.99	41.90	58.10	63.62	80.22	77.20	1759
C								
C_1	61.19	14.92	21.46	71.53	59.68	53.06	50.23	657
C_2	70.46	18.24	32.50	68.51	59.39	67.42	63.84	1283
C_3	74.81	24.27	37.56	60.74	60.12	76.20	72.02	647
C_4	86.02	38.06	58.25	39.03	66.80	87.18	86.60	515
AB								
A_1B_1	58.10	7.22	17.53	66.32	43.64	43.99	42.26	291
A_1B_2	50.60	5.33	16.27	80.77	42.90	41.72	40.23	338
A_1B_3	76.50	23.58	34.59	67.61	65.72	72.64	68.84	318
A_2B_1	68.80	20.54	27.68	51.78	61.61	61.61	56.25	112
A_2B_2	75.20	18.26	36.51	68.88	63.49	65.15	61.82	241
A_2B_3	77.40	28.59	40.47	57.03	62.66	79.84	76.56	640
A_3B_1	84.60	18.31	45.07	54.93	61.97	73.24	70.42	71
A_3B_2	74.20	19.31	36.90	66.20	58.97	65.86	62.75	290
A_3B_3	77.70	31.46	45.94	55.18	63.55	83.52	81.02	801
AC								
A_1C_1	59.37	11.61	19.24	72.96	48.92	48.42	46.43	603
A_1C_2	59.76	6.51	23.67	77.51	46.15	47.34	43.78	169
A_1C_3	64.29	19.31	23.47	68.36	47.14	63.27	60.20	98
A_1C_4	79.22	18.18	48.05	55.84	67.53	85.71	84.41	77
A_2C_1	83.33	61.11	61.11	50.00	72.22	94.46	94.44	18
A_2C_2	80.29	16.60	31.76	67.00	60.86	65.37	60.45	488
A_2C_3	76.21	24.16	36.43	61.71	62.08	79.55	67.48	269
A_2C_4	87.16	42.64	52.29	39.90	66.97	85.78	85.78	218
A_3C_1	80.56	47.22	38.89	58.33	69.44	91.67	91.66	36
A_3C_2	73.48	22.68	35.46	67.25	61.82	74.44	71.88	626
A_3C_3	77.14	26.07	43.57	57.14	59.29	77.50	72.85	280
A_3C_4	87.27	40.45	67.73	32.37	66.36	89.09	88.18	220
BC								
B_1C_1	56.84	5.98	14.53	67.52	42.31	41.88	40.59	234
B_1C_2	66.67	5.67	25.43	64.54	51.77	49.65	43.93	141
B_1C_3	70.21	25.53	34.04	53.19	65.96	74.47	70.21	47
B_1C_4	88.46	44.23	53.85	30.77	71.15	88.46	88.46	52
B_2C_1	46.97	7.58	15.66	81.81	42.42	42.93	41.41	198
B_2C_2	66.50	8.68	27.79	78.16	54.59	55.09	52.35	403
B_2C_3	68.24	13.51	27.03	68.24	48.65	56.76	51.35	148
B_2C_4	87.50	40.00	55.83	44.16	77.50	80.67	81.66	120
B_3C_1	78.22	30.67	33.78	66.66	67.67	70.67	68.00	225
B_3C_2	73.34	25.85	36.40	64.00	63.46	77.54	73.88	739
B_3C_3	77.43	27.65	41.37	59.07	63.27	82.74	78.98	452
B_3C_4	85.13	36.44	59.77	38.48	62.39	88.92	88.04	343

1950-1960 Growth Concentration

Manufacturing Change Variable	All Establish-ments	Large Establish-ments	Medium Size Establish-ments	Employment Place of Residence	Employment Place of Work	Payroll	Value-Added	Number of Counties
a								
a_1+a_2	70.03	19.62	31.14	62.52	54.26	65.78	63.52	1198
a_3	74.94	21.61	36.32	65.28	60.69	73.91	70.80	870
a_4	71.76	25.34	39.85	60.54	62.86	69.44	66.15	1034
b								
b_1	71.63	21.76	34.80	62.27	57.82	69.97	67.14	2767
b_2	74.92	24.78	41.19	65.67	68.06	63.58	60.59	335
C								
C_1	61.19	14.92	21.46	71.53	59.68	53.06	50.23	657
C_2	70.46	18.24	32.50	68.51	59.39	67.42	63.84	1283
C_3	74.81	24.27	37.56	60.74	60.12	76.20	72.02	647
C_4	86.02	38.06	58.25	39.03	66.80	87.18	86.60	515
ab								
a_1b_1	60.90	13.65	20.07	70.67	49.90	50.71	48.88	491
a_1b_2	55.70	4.35	15.65	80.00	46.96	40.87	38.26	115
a_2b_1	80.20	27.24	42.00	52.72	58.17	82.78	80.14	569
a_2b_2	87.00	34.78	69.57	43.47	86.96	91.30	91.30	23
a_3b_1	74.40	20.98	36.07	64.78	59.63	74.11	71.04	815
a_3b_2	83.70	30.91	40.00	72.72	76.36	70.81	67.27	55
a_4b_1	69.70	23.43	37.00	61.43	60.39	68.61	65.35	892
a_4b_2	85.30	37.32	57.75	54.93	78.87	74.65	71.12	142
aC								
$a_1+a_2C_{1+2}$	66.05	16.10	25.46	69.64	53.75	60.17	57.67	919
$a_1+a_2C_3$	81.88	26.17	41.61	48.99	58.39	79.20	77.18	149
$a_1+a_2C_4$	74.62	36.92	59.23	27.69	53.08	90.00	89.23	130
a_3C_{1+2}	71.66	18.97	31.85	71.42	60.42	66.28	62.53	427
a_3C_3	71.49	18.88	32.93	67.87	59.44	77.71	72.69	249
a_3C_4	86.60	30.93	50.52	48.87	62.89	87.11	86.59	194
a_4C_{1+2}	66.16	17.34	31.65	67.84	57.91	62.63	59.26	594
a C	73.90	28.51	39.76	61.04	61.45	73.49	68.27	249
a C	86.39	46.07	65.45	37.14	80.10	85.34	84.81	191
bC								
b_1C_1	62.06	16.82	22.24	69.72	51.22	54.02	52.33	535
b_1C_2	70.03	18.86	32.14	67.69	58.93	68.11	64.52	1198
b_1C_3	74.63	23.55	37.31	60.20	58.97	76.45	72.63	603
b_1C_4	83.76	33.41	54.29	40.83	61.48	85.85	85.15	431
b_2C_1	57.38	6.56	18.03	78.68	49.18	43.44	40.98	122
b_2C_2	76.47	9.41	37.65	80.00	65.88	58.82	54.11	85
b_2C_3	77.23	34.09	40.91	71.45	75.00	70.45	63.63	44
b_2C_4	97.56	61.90	78.57	29.76	94.05	94.05	94.04	84

in terms of industrial employment by place of residence. If that trend continues, more industrial workers will be located, relatively, in peripheral areas than in core regions. Such a trend conforms to the shift to tertiary and higher sectors of employment in the core urban areas of the country, and the entrance of rural population into the manufacturing labor force all across the country.

Although the generalizations stated in the above paragraphs are true, one should also note the differences among the industrial change indicators. Clearly, the spatial processes of diffusion, dispersion and decentralization did not hold as strongly with respect to change in the number of large establishments as with change in the number of medium size establishments or with change in all manufacturing establishments. This is an indication that in spite of diffusion, decentralization and dispersion, the core areas of the country were still the most important ones and that the periphery, although reduced, was making at best a modest attempt to catch up, particularly through the spread of smaller industrial establishments. Such catching up was indeed, more a result of the filtering up of countries to higher categories of relative location than it was of growth acceleration markedly in the peripheries.

What has been said so far is meant to put in perspective the operation of hierarchical diffusion, dispersion and decentralization. The conclusion becomes obvious if one notes the high degree of similarity between subtables 1 and 2 in Table 7.1. The two subtables represent industrial change according to relative location in 1950 and 1960, respectively.

We have indicated the deviation of the change in manufacturing by place of residence from all other variables. It is important however, to contrast it with the change in employment by place of work. The growth probabilities of jobs in manufacturing are higher in core areas than in peripheral ones, an indication that during the 1950's the location of manufacturing establishments was still determined primarily by centrality and accessibility and was dependent on the economies that could be derived from locating at focal points throughout the Nation's economic space, whereas an increasingly sidespread distribution of places of residence was emerging.

The last group of industrial change variables confirm that although diffusion, dispersion and decentralization processes had their impact on peripheral areas, the peripheral areas were only holding the line in the period studied. The probability of counties experiencing growth is payroll or in value added was much higher in the core areas than in peripheral ones. This lends support to the hypothesis of circular and cumulative causation. What we see is the entire system being elevated to a new level--the dominance of the old manufacturing belt was reduced by the spread of the heartland.

The third subtable in Table 7.1 indicates the extent of change in the manufacturing variables with respect to change in the components of relative location. With few exceptions, the results indicate the significance of spatial dynamism. The greater the change in central city populations and in the relative accessibility of counties, the higher the growth probability for any industrial change variable. A complicating factor is the relative location of counties within urban fields. However, this variable indicates the significance of proximity to central city regardless of change in other spatial components. The closer a county is to a central city of an economic area the higher its growth probability regardless of change in city size or in accessibility.

The second summary table, Table 7.2 draws together the results of the contingency analyses. The first column for each manufacturing change variable shows the proportion of the total variation which is due to growth in each relative location category. The second column

TABLE 7.2

Summary of Contingency Analysis

Association Between Relative Location Components and Manufacturing Change Variables

Growth Category Only: 1950, 1960 and 1950-1960

1950

Manufacturing Change Variable	All Establishments χ^2	Direction	Large Establishments χ^2	Direction	Medium Size Establishments χ^2	Direction
A Total Chi Square	75.15 Sig. <.001		67.17 Sig. <.001		96.91 Sig. <.001	
A_1	16.62	—	51.02	—	38.55	—
A_2	.87	+	6.77	+	2.34	+
A_3	10.52	+	22.36	+	23.73	+
B	81.95 Sig. <.001		102.94 Sig. <.001		93.60 Sig. <.001	
B_1	14.89	—	34.97	—	35.45	—
B_2	.00	—	3.87	—	.75	—
B_3	12.67	+	39.07	+	28.31	+
C	93.23 Sig. <.001		107.93 Sig. <.001		178.36 Sig. <.001	
C_1	11.43	—	14.11	—	20.37	—
C_2	.46	—	7.86	—	1.78	—
C_3	.74	+	1.27	+	.44	+
C_4	15.45	+	54.64	+	41.94	+
AB	140.69 Sig. <.001		132.67 Sig. <.001		147.08 Sig. <.001	
A_1B_1	17.05	—	38.21	—	35.27	—
A_1B_2	1.24	—	8.12	—	2.18	—
A_1B_3	.98	+	2.55	+	.14	—
A_2B_1	.09	—	.18	—	.56	—
A_2B_2	.06	—	.00		.42	—
A_2B_3	1.97	+	9.22	+	7.70	+
A_3B_1	1.83	+	.00		.88	+
A_3B_2	.26	+	.04	—	.90	+
A_3B_3	4.58	+	19.58	+	16.55	+
AC	127.43 Sig. <.001		179.97 Sig. <.001		234.95 Sig. <.001	
A_1C_1	10.44	—	16.46	—	19.01	—
A_1C_2	2.83	—	12.10	—	3.06	—
A_1C_3	.71	—	.08	—	2.37	—
A_1C_4	.51	+	1.13	+	4.04	+
A_2C_1	.24	+	1.48	+	1.14	+
A_2C_2	.14	—	2.62	—	1.04	—
A_2C_3	.49	+	.75	+	.43	+
A_2C_4	5.45	+	18.52	+	7.16	+
A_3C_1	.27	+	5.62	+	.14	+
A_3C_2	.14	+	.22	+	.10	+
A_3C_3	.76	+	.51	+	1.79	+
A_3C_4	6.12	+	17.56	+	24.06	+
BC	158.51 Sig. <.001		294.54 Sig. <.001		246.35 Sig. <.001	
B_1C_1	9.32	—	14.90	—	20.22	—
B_1C_2	2.02	—	10.37	—	5.58	—
B_1C_3	.09	—	.07	—	.34	—
B_1C_4	2.31	+	7.82	+	5.78	+
B_2C_1	.59	—	.84	—	1.43	—
B_2C_2	.59	—	3.92	—	.54	—
B_2C_3	.20	—	1.36	—	.62	—
B_2C_4	2.21	+	8.32	+	4.58	+
B_3C_1	.74	+	3.82	+	.01	—
B_3C_2	.83	+	20.29	+	.25	+
B_3C_3	2.16	+	4.07	+	2.82	+
B_3C_4	4.11	+	6.16	+	22.53	+

Employment: Place of Residence χ^2	Direction	Employment: Place of Work χ^2	Direction	Payroll χ^2	Direction	Value-Added χ^2	Direction
39.14	Sig. <.001	30.08	Sig. <.001	180.72	Sig. <.001	162.45	Sig. <.001
24.45	+	26.63	—	19.14	—	20.30	—
5.49	—	4.65	+	1.88	+	1.38	+
7.43	—	9.81	+	9.70	+	11.87	+
65.97	Sig. <.001	39.47	Sig. <.001	248.58	Sig. <.001	230.42	Sig. <.001
1.53	+	27.51	—	18.44	—	19.70	—
15.87	+	.15	+	.06	—	.13	—
19.87	—	13.43	+	12.21	+	13.73	+
165.91	Sig. <.001	31.70	Sig. <.001	185.36	Sig. <.001	185.18	Sig. <.001
5.10	+	23.75	—	15.14	—	14.15	—
4.22	+	.16	+	.35	—	.69	—
.22	—	.54	+	2.44	+	1.62	+
27.77	—	16.59	+	12.80	+	17.08	+
90.33	Sig. <.001	81.76	Sig. <.001	322.04	Sig. <.001	294.93	Sig. <.001
1.85	+	27.76	—	18.47	—	19.52	—
14.67	+	1.52	—	.34	—	1.22	—
.20	—	6.21	+	.34	+	.44	+
1.04	—	.11	+	.77	—	.98	—
1.74	+	.95	+	.15	+	.02	+
7.76	—	.73	+	2.81	+	2.98	+
.07	+	.95	+	.04	+	.08	+
.46	+	.55	+	.18	+	.20	+
8.54	—	2.32	+	6.99	+	8.10	+
189.71	Sig. <.001	54.90	Sig. <.001	307.08	Sig. <.001	306.12	Sig. <.001
5.34	+	17.81	—	12.14	—	11.70	—
2.72	+	7.89	—	4.98	—	5.45	—
.23	+	.00		.09	—	.06	—
1.74	—	4.10	+	1.57	+	2.00	+
.05	—	.72	+	.24	+	.47	+
.95	+	.38	+	.24	—	.71	—
.52	—	.27	+	1.74	+	1.12	+
8.50	—	3.20	+	1.94	+	3.08	+
.09	—	.60	+	.83	+	1.42	+
.96	+	2.71	+	1.44	+	1.66	+
.13	—	.05	+	.62	+	.46	+
16.08	—	3.22	+	4.85	+	5.53	+
219.23	Sig. <.001	123.83	Sig. <.001	379.58	Sig. <.001	366.96	Sig. <.001
2.22	+	18.86	—	12.56	—	12.71	—
1.16	+	4.30	—	5.85	—	6.72	—
.00		.06	+	.18	—	.31	—
6.72	—	7.49	+	1.42	+	2.06	+
3.67	+	.00		.42	—	.39	—
6.40	+	.10	+	.09	—	.19	—
.47	+	1.78	—	.01	—	1.26	—
3.17	—	2.78	+	.86	+	.18	+
.08	—	1.62	+	.13	+	.12	+
.13	—	2.09	+	1.72	+	1.26	+
1.04	—	1.95	+	3.26	+	3.12	+
12.26	—	.01	+	4.00	+	5.98	+

1960

Manufacturing Change Variable	All Establishments Percent Variation	All Establishments Chi-Square Deviation	Large Establishments Percent Variation	Large Establishments Chi-Square Deviation	Medium Size Establishments Percent Variation	Medium Size Establishments Chi-Square Deviation
A Total Chi-Square	74.53 Sig. <.001		81.50 Sig. <.001		103.11 Sig. <.001	
A_1	19.28	—	52.98	—	41.67	—
A_2	2.86	+	5.39	+	1.71	+
A_3	5.97	+	19.56	+	21.24	+
B	57.58 Sig. <.001		114.01 Sig. <.001		74.94 Sig. <.001	
B_1	6.23	—	19.24	—	23.16	—
B_2	9.66	—	25.01	—	14.57	—
B_3	12.12	+	33.65	+	26.78	+
C	93.23 Sig. <.001		107.93 Sig. <.001		178.36 Sig. <.001	
C_1	11.43	—	14.11	—	20.36	—
C_2	.46	—	7.86	—	1.78	—
C_3	.74	+	1.27	+	.41	+
C_4	15.44	+	54.65	+	41.94	+
AB	135.63 Sig. <.001		154.17 Sig. <.001		147.29 Sig. <.001	
A_1B_1	5.64	—	18.80	—	17.82	—
A_1B_2	15.73	—	28.03	—	23.90	—
A_1B_3	.63	+	.23	+	.05	—
A_2B_1	.15	—	.10	—	1.37	—
A_2B_2	.26	+	.99	—	.03	+
A_2B_3	1.85	+	8.11	+	3.06	+
A_3B_1	1.17	—	.36	—	1.33	+
A_3B_2	1.25	+	.65	—	.10	+
A_3B_3	2.54	+	20.61	+	16.86	+
AC	136.73 Sig. <.001		199.86 Sig. <.001		230.70 Sig. <.001	
A_1C_1	9.95	—	14.93	—	19.45	—
A_1C_2	3.52	—	9.14	—	2.89	—
A_1C_3	.28	—	.20	—	1.78	—
A_1C_4	1.20	+	.26	—	1.60	+
A_2C_1	.24	+	6.12	+	1.81	+
A_2C_2	.28	—	3.38	—	.81	—
A_2C_3	.10	+	.30	+	.04	+
A_2C_4	3.39	+	21.11	+	7.70	+
A_3C_1	.47	+	5.06	+	.03	+
A_3C_1	.74	+	.06	+	.00	
A_3C_3	1.51	+	.98	+	2.31	+
A_3C_4	6.26	+	16.34	+	25.79	+
BC	161.86 Sig. <.001		234.30 Sig. <.001		216.71 Sig. <.001	
B_1C_1	4.50	—	11.85	—	13.34	—
B_1C_2	.39	—	7.32	—	1.80	—
B_1C_3	.01	—	.17	+	.03	—
B_1C_4	1.35	+	4.30	+	2.57	+
B_2C_1	10.80	—	8.16	—	10.03	—
B_2C_2	1.03	—	13.98	—	3.10	—
B_2C_3	.21	—	2.18	—	1.47	—
B_2C_4	2.59	+	7.95	+	6.18	+
B_3C_1	.75	+	3.08	+	.09	—
B_3C_2	.12	+	2.05	+	.09	+
B_3C_3	1.19	+	2.67	+	2.10	+
B_3C_4	5.07	+	14.22	+	22.38	+

Employment: Place of Residence Percent Variation	Chi-Square Deviation	Employment: Place of Work Percent Variation	Chi-Square Deviation	Payroll Percent Variation	Chi-Square Deviation	Value-Added Percent Variation	Chi-Square Deviation
50.76 Sig. <.001		37.28 Sig. <.001		178.38 Sig. <.001		161.77 Sig. <.001	
25.72	+	28.49	—	20.80	—	22.40	—
3.44	—	6.62	+	1.96	+	1.65	+
8.19	—	5.95	+	7.97	+	9.51	+
53.19 Sig. <.001		37.83 Sig. <.001		229.26 Sig. <.001		212.57 Sig. <.001	
.30	—	14.41	—	8.30	—	9.31	—
26.15	+	9.54	—	9.25	—	9.86	—
10.92	—	17.13	+	13.19	+	14.38	+
165.91 Sig. <.001		31.70 Sig. <.001		185.36 Sig. <.001		185.18 Sig. <.001	
5.10	+	23.75	—	15.14	—	14.15	—
4.21	+	.16	+	.35	—	.69	—
.22	—	.54	+	2.44	+	1.62	+
27.77	—	16.59	+	12.80	+	17.08	+
92.61 Sig. <.001		104.77 Sig. <.001		328.28 Sig. <.001		296.16 Sig. <.001	
.71	+	17.94	—	8.26	—	8.57	—
18.95	+	23.19	—	11.26	—	11.88	—
1.39	+	4.08	+	.17	+	.10	+
2.22	—	.21	+	.32	—	.55	—
1.61	+	1.35	+	.18	—	.26	—
3.49	—	2.42	+	3.18	+	3.36	+
.62	—	.15	+	.05	+	.06	+
.60	+	.00		.15	—	.21	—
7.74	—	4.35	+	7.13	+	8.53	+
196.88 Sig. <.001		57.43 Sig. <.001		289.74 Sig. <.001		290.36 Sig. <.001	
5.16	+	17.65	—	13.10	—	12.57	—
3.00	+	8.43	—	4.04	—	4.44	—
.30	+	.12	—	.18	—	.19	—
.26	—	1.90	+	1.10	+	1.32	+
.18	—	.62	+	.72	+	.72	+
.73	+	.49	+	.37	—	.90	—
.01	—	.70	+	1.46	+	1.11	+
9.27	—	4.40	+	2.96	+	4.19	+
.09	—	1.32	+	.88	+	1.16	+
1.09	+	1.52	+	.81	+	.96	+
.66	—	.01	+	.94	+	.58	+
16.52	—	3.90	+	4.17	+	5.43	+
210.51 Sig. <.001		99.21 Sig. <.001		367.80 Sig. <.001		372.22 Sig. <.001	
.39	+	11.10	—	7.33	—	6.49	—
.05	+	1.21	—	2.18	—	2.93	—
.26	—	.32	+	.03	+	3.47	+
4.16	—	1.17	+	.76	+	.93	+
5.53	+	9.38	—	5.37	—	5.09	—
7.48	+	1.22	—	3.16	—	3.26	—
.33	+	2.61	—	.95	—	1.33	—
3.06	—	6.87	+	.74	+	1.09	+
.27	+	2.19	+	.01	+	.03	+
.10	+	2.52	+	1.98	+	1.65	+
.43	—	1.51	+	3.12	+	2.91	+
15.22	—	.85	+	5.48	+	6.66	+

1950-1960

Manufacturing Change Variable	All Establishments Percent Variation	All Establishments Chi-Square Deviation	Large Establishments Percent Variation	Large Establishments Chi-Square Deviation	Medium Size Establishments Percent Variation	Medium Size Establishments Chi-Square Deviation
a Total Chi-Square	6.07 Sig. <.05		10.97 Sig. <.10		18.67 Sig. <.001	
a_1+a_2	10.04	—	30.99	—	34.06	—
a_3	17.79	+	.73	—	.86	+
a_4	.16	—	46.30	+	29.57	+
b	1.65 Sig. <.05		1.57 Sig. <.30		5.27 Sig. <.020	
b_1	3.03	—	8.28	—	7.02	—
b_2	24.84	+	69.42	+	57.49	+
C	93.23 Sig. <.001		107.93 Sig. <.001		178.36 Sig. <.001	
C_1	11.43	—	14.11	—	20.36	—
C_2	.46	—	7.86	—	1.78	—
C_3	.74	+	1.27	+	.41	+
C_4	15.44	+	54.65	+	41.49	+
ab	87.12 Sig. <.001		73.23 Sig. <.001		122.73 Sig. <.001	
a_1b_1	9.48	—	21.25	—	25.64	—
a_1b_2	4.95	—	21.85	—	10.51	—
a_2b_1	5.90	+	9.11	+	5.52	+
a_2b_2	.60	+	2.46	+	6.52	+
a_3b_1	.71	+	.61	—	.05	+
a_3b_2	1.03	+	2.84	+	.16	+
a_4b_1	.78	—	1.00	+	.43	—
a_4b_2	4.54	+	18.82	+	15.98	+
aC	100.04 Sig. <.001		128.48 Sig. <.001		177.30 Sig. <.001	
$a_1+a_2C_{1+2}$	4.57	—	11.60	—	14.64	—
$a_1+a_2C_3$	2.10	+	.84	+	.86	+
$a_1+a_2C_4$	2.72	+	9.69	+	11.75	+
a_3C_{1+2}	.00	—	1.40	—	.95	—
a_3C_3	.00	—	.90	—	.23	—
a_3C_4	5.60	+	5.23	+	6.88	+
a_4C_{1+2}	2.86	—	4.65	—	1.42	—
a_4C_3	.14	+	3.62	+	.78	+
a_4C_4	5.72	+	39.21	+	26.95	+
bC	101.18 Sig. <.001		166.50 Sig. <.001		210.90 Sig. <.001	
b_1C_1	7.21	—	3.99	—	12.58	—
b_1C_2	.60	—	3.45	—	1.78	—
b_1C_3	.58	+	.37	+	.27	+
b_1C_4	8.29	+	14.41	+	20.33	+
b_2C_1	3.64	—	8.03	—	5.78	—
b_2C_2	.26	+	3.82	—	.06	+
b_2C_3	.12	+	1.50	+	.11	+
b_2C_4	7.15	+	43.28	+	24.45	+

Employment: Place of Residence		Employment: Place of Work		Payroll		Value-Added	
Percent Variation	Chi-Square Deviation	Percent Variation	Chi-Square Deviation	Percent Variation	Chi-Square Deviation	Percent Variation	Chi-Square Deviation
4.60 Sig. <.200		18.60 Sig. <.001		15.57 Sig. <.001		12.08 Sig. <.010	
.00	—	23.87	—	13.62	—	12.75	—
21.08	+	2.36	+	17.02	+	20.70	+
16.30	—	14.84	+	.06	+	.08	—
1.44 Sig. <.200		13.28 Sig. <.001		5.67 Sig. <.020		6.01 Sig. <.020	
4.16	—	4.44	—	3.35	+	3.66	+
33.33	+	36.75	+	27.51	—	29.78	—
165.91 Sig. <.001		31.70 Sig. <.001		185.36 Sig. <.001		185.18 Sig. <.001	
5.10	+	23.75	—	15.14	—	14.15	—
4.21	+	.16	+	.35	—	.69	—
.22	—	.54	+	2.44	+	1.62	+
27.77	—	16.59	+	12.80	+	17.08	+
61.61 Sig. <.001		60.99 Sig. <.001		189.13 Sig. <.001		171.68 Sig. <.001	
8.01	+	10.66	—	12.88	—	13.22	—
9.02	+	4.72	—	7.20	—	7.85	—
14.30	—	.08	—	7.96	+	9.38	+
1.85	—	4.21	+	.82	+	1.40	+
.92	+	.13	+	1.42	+	1.56	+
1.15	+	5.11	+	.02	+	.00	
.36	—	.44	+	.03	—	.10	—
1.85	—	15.30	+	.34	+	.22	+
192.48 Sig. <.001		52.94 Sig. <.001		146.50 Sig. <.001		157.98 Sig. <.001	
3.69	+	8.03	—	7.56	—	6.80	—
2.23	—	.02	—	1.49	+	1.64	+
12.99	—	1.57	—	5.53	+	6.62	+
2.81	+	.26	+	.39	—	.65	—
.56	+	.02	+	1.28	+	.98	+
3.34	—	1.06	+	6.24	+	7.46	+
1.34	+	.19	—	2.64	—	2.96	—
.05	—	.45	+	3.96	+	.06	+
10.40	—	30.01	+	5.34	+	6.51	+
180.57 Sig. <.001		66.92 Sig. <.001		200.09 Sig. <.001		197.60 Sig. <.001	
2.39	+	7.98	—	9.06	—	8.02	—
2.75	+	.00		.12	—	.33	—
.33	—	.00		2.21	+	1.73	+
18.13	—	.72	+	8.43	+	11.61	+
3.12	+	2.99	—	6.02	—	6.00	—
2.34	+	1.08	+	.68	—	1.07	—
.18	+	2.81	+	.01	+	.01	—
8.19	—	25.13	+	4.24	+	4.78	+

shows the deviation of growth from expected proportions based upon the marginals. A minus sign indicates a less-than-expected growth proportion and a plus sign indicates a larger-than-expected growth.

Strikingly, all manufacturing change variables for all relative location categories show similar and consistent deviation patterns. The only exception is the change in manufacturing employment by place of residence, where the pattern is reversed.

The chi-square analysis thus reveals the following:

A. With respect to hierarchical diffusion --

1. Industrial growth in the 1950's was positively related to location within the economic sphere of high status centers at the beginning of the decade. The same results hold for 1960 (see subtable 2 Table 7.2). While many counties were improving their relative location, the positive relationships between hierarchical status and industrial growth prospects were preserved during the decade, meaning that there was a growing number of counties within the status range most likely to benefit from industrial growth.
2. Rather than accelerated growth filtering down, counties "filtered up." Growth prospects did not increase for lower-status counties; instead counties' growth prospects increased as the economic centers of their FEA's changed in hierarchical status.
3. With respect to the similarities and differences among the industrial change variables, the hierarchical diffusion effect is strong for the change in the total number of manufacturing establishments. It is most extensive for small-scale plants and least extensive for large plants. Thus, the city-size effect on the location of new industrial plants approaches insignificance for small plants. Counties in urban fields where the central city grew very rapidly (see subtable 3, Table 7.2) had much greater growth probabilities than most other cases.
4. The ironical result is that continuous economic growth and substantial diffusion went hand in hand with increasing concentration of economic activity in the largest urban regions. It was because the number of such regions was growing that the periphery was reduced.

B. With respect to dispersion --

1. Manufacturing growth favored areas of high national market access, especially those whose national market access is increasing. Thus, substantial support is given to Harris's hypothesis of the importance of relative accessibility in the location of manufacturing activities.

2. Not only were counties able to improve their growth prospects by improving their national market access; industrial growth was apparently slackening somewhat in the highest-access areas, while still favoring them relatively, and increasing due to dispersion into peripheral areas.

3. A trend toward the dispersion of industrial plants and workers as well as of value added and payroll was evident. This trend was evident for 1950 as well as for 1960, after other changes in the spatial components of relative location and the filtering up process were taken into account.

4. The increase in relative accessibility throughout the system did not change the extent of dispersion, meaning that many counties were doubly blessed by extension of the heartland as well as by dispersion. (Compare subtable 1 with 2 in Table 7.2).

C. With respect to decentralization --

1. Decentralization was clearly taking place, and was highly selective towards the further peripheries.

2. There was extensive decentralization of place of residence during the 1950's, involving substantial increases in long-distance commuting.

3. The familiar decay from the center of economic areas towards their peripheries was present for all manufacturing change variables. It was most explicit in the case of value added by manufacturing.

D. With respect to the joint effects of hierarchical status and dispersion

1. Growth prospects were greater than expected for high status FEA's throughout the country in 1950 and in 1960, and for all FEA's in the zone of highest national market access in those years.

2. The reduction in the geographical extent of the periphery was clearly apparent.

3. The cause for negative deviations in the periphery both in 1950 and in 1960 was lack of growth of the centers, regardless of the change in accessibility. Thus, a growth center policy that relies only on changing accessibility is not, as many studies demonstrate, sufficient. There must be a significant change in the size of the centers of economic areas. Generally, the greater the change the better.

4. In spite of simultaneous hierarchical diffusion and dispersion, groups A_2B_3 and A_3B_3, the relative location cores of the country, were preferred by newly-located industrial establishments as well as by all other manufacturing variables.

5. If an FEA center showed a dynamic character, associated counties were more ripe for change in all manufacturing variables.

E. With respect to the joint effects of hierarchical status and decentralization

1. Most negative deviations from expectations were concentrated in the outlying counties in the urban fields of small urban centers.

2. The most pronounced aspect of decentralization was the process of suburbanization at all levels of the urban hierarchy. The higher the level of the urban hierarchy, the more extensive and accelerated the process of suburbanization, extending much further out into the largest urban fields. This was true for 1950 as well as for 1960. The change in hierarchical status also was associated with manufacturing change, to the exclusion of employment change, in a similar manner: the greater the change in the size of the FEA city the greater and more extensive the decentralization process.

3. Large manufacturing plants tended to locate in the largest economic areas, but close to the outer margin of daily commuting.

4. Decentralization took place regardless of the degree of change in the FEA city population. On the other hand, positive deviations from the expected proportions were always associated with areas of intensive commuting and negative deviations with marginal location in urban fields.

5. One suspects that relative location of counties had a more selective effect on the change in the number of manufacturing plants than it had on employment change by place of work because many more factors unrelated to relative location influence the change in employment than the change in number of establishments.

F. With respect to the joint effects of dispersion and decentralization

1. A clear pattern emerges: industrial dispersion from areas of high relative accessibility to areas of low relative accessibility was affected by relative location within urban fields. For any level of relative accessibility, the closer counties were to their urban centers, the greater was the likelihood of growth. This was true for relative location in 1950 as well as 1960 and for all manufacturing variables, excluding the change in employment by place of residence.

2. Rapid change in relative accessibility affected change positively except at the periphery of urban regions.

3. Generally, by 1960 the importance of the B_iC_i interaction declined as compared to the levels of the 1950s. It appears that the interaction between the relative location components was not a very powerful as a source of deviations from the national proportion of counties experiencing growth and decline.

4. Improved accessibility resulted in a relatively more equitable distribution of the proportion of counties that experienced growth in areas with low and medium levels of relative accessibility than would have been the case if there had been little or no change in relative accessibility during the 1950's.

Relative location and rate of change differentials

A summary of the analysis of variance of the association of relative location and rate of industrial change differentials is given in Table 7.3. The four subtables represent the results for 1950, 1960 for the change between 1950 and 1960, and for the initial relative location and its change. The following summarizes the most important conclusions:

A. For 1950 and 1960

1. Given relative location in 1950, the most important elements of location of establishments and workers in the national space-economy were relative market access together with its interaction with the location of counties within urban fields.

2. The most consistently significant and general source of change and growth rate differentials in the decade was national market access, thus confirming Harris's speculations. Hierarchical status and local market access tended to be factors whose significance varied from one zone of national market access to another, and therefore assumed significance only in their interaction with the more general variable.

3. For all establishments, the dominant process seems to be that of dispersion. The inference can be made that hierarchical diffusion and decentralization are more derivative than independent processes. There are several reasons why this may be the case. One answer could be that the smaller the plant, the greater the reliance on relative accessibility and the less the weight of other factors. As plant size increases, accessibility remains significant, but it is not the only determinant. City size is irrelevant for small plants but becomes increasingly important with increases in plant size. The significance

TABLE 7.3

Summary of Analysis of Variance: Most Significant Effects

1950

Manufacturing Change Variable	All Establish-ments	Large Establish-ments	Medium Size Establish-ments	Employment Place of Residence	Employment Place of Work	Payroll	Value-Added
Effects							
	BC	B	B	B	BC	C	A
	B	A	A(?)	A	AC	BC	C
		C		C		AB	AC
		BC		AB		AC	AB
						A	
Mean Squares							
	31.40	72.21	27.73	11852.21	30.46	97.66	93.11
	57.89	10.47	10.20	5213.09	13.28	81.45	62.56
		7.39		3278.44		52.68	36.26
		4.70		886.73		56.19	78.12
						65.29	87.37
Degrees of Freedom							
	6,3066	2,3066	2,3066	2,3066	3,3066	3,3066	2,3066
	2,3066	2,3066	2,3066	2,3066	6,3066	6,3066	3,3066
		3,3066		3,3066		4,3066	6,3066
		6,3066		4,3066		6,3066	4,3066
						2,3066	
F-Ratio							
	3.24	51.26	5.76	44.70	4.84	7.51	5.67
	5.98	7.42	2.12	19.66	2.11	6.26	3.81
		5.24		12.37		4.05	2.21
		3.33		3.34		4.32	4.76
						5.02	
Significance							
	<.0035	<.0001	<.0032	<.0001	<.0001	<.0001	<.0035
	<.0026	<.0007	<.1199	<.0001	<.0487	<.0001	<.0096
		<.0014		<.0001		<.0028	<.0391
		<.0029		<.0097		<.0003	<.0008
						<.0067	
Least Square Estimates							
	-2.05	-.35	-.24	+4.08	-.89	-.08	-.74
	- .03	-.25	-.17	+3.33	+.32	-1.19	-.35
	-1.51	-.20	-.19	+4.41	-1.04	-1.08	+.18
	- .17	.00	-.01	+1.54	-.75	-1.64	-.86
	-1.44	.25		+4.57	-.64	+.57	-.46
	.08	.34		+5.15	-.63	-2.22	-1.06
	.42	.26		+3.20	-.32	-.58	+2.10
	- .14	-.58		+4.84	-.45	-3.50	-.68
		-.42		+2.61	+.16	-1.06	+.58
		-.57		-1.61	+.26	-2.36	+.20
		-.41		-2.24	+.70	-1.93	+.76
		-.25			+.36	-1.00	-2.05
		-.41				-.50	-1.31
						-1.15	-1.63
						+.91	-.75
						+.19	
						+.76	
						+1.73	
						+1.26	
						-.75	
						-.58	

1960

Manufacturing Change Variable	All Establishments	Large Establishments	Medium Size Establishments	Employment Place of Residence	Employment Place of Work	Payroll	Value-Added
Effects							
	BC	B	B	A	AC	C	C
	AB(?)	A	A	B	BC	BC	BC
		C		C		AB	A
		BC		AB		AC	B
						A	
Mean Squares							
	30.23	72.89	23.91	12878.47	20.84	97.66	87.37
	15.26	14.18	15.04	6014.57	17.02	57.17	61.84
		7.06		3070.30		49.94	65.07
		3.79		882.74		39.84	39.38
						36.83	
Degrees of Freedom							
	6,3066	2,3066	2,3066	2,3066	6,3066	3,3066	3,3066
	4,3066	2,3066	2,3066	2,3066	6,3066	6,3066	6,3066
		3,3066		3,3066		4,3066	2,3066
		6,3066		4,3066		6,3066	2,3066
						2,3066	
F-Ratio							
	3.09	51.73	4.97	N.A.	3.29	7.42	5.34
	1.56	10.07	3.12		2.69	4.35	3.78
		5.01				3.80	3.98
		2.69				3.03	2.41
						2.80	
Significance							
	<.0051	<.0001	<.0070	<.0001	<.0031	<.0001	<.0012
	<.1805	<.0001	<.0439	<.0001	<.0131	<.0003	<.0010
		<.0019		<.0001		<.0044	<.0187
		<.0131		<.0098		<.0059	<.0395
						<.0607	
Least Square Estimates							
	N.A.	-.27	-.24	+5.50	-.55	-.14	-.76
		-.33	-.02	+2.07	+.43	-1.21	-1.04
		-.23	-.08	+2.39	-.12	-.01	-.43
		.01	-.23	+4.09	+.33	-1.26	-1.10
		-.29		+5.24	+.64	-1.24	-2.17
		-.38		+5.55	+.50	-1.03	-1.07
		-.28		+4.20	-.75	-1.79	-1.58
		-.41		+4.71	-1.04	-1.56	.44
		-.51		+7.41	-.12	-2.33	-1.97
		-.48		+.82	-.99	-.18	-.64
		-.48		+4.27	-.30	-1.81	-.20
		-.15		+2.82	-.82	.14	-.24
		-.32		+3.12		-.65	-.32
						-1.02	
						1.37	
						.18	
						.64	
						1.47	
						1.10	
						-.60	
						-.09	

1950-1960

Manufacturing Change Variable	All Establishments	Large Establishments	Medium Size Establishments	Employment Place of Residence	Employment Place of Work	Payroll	Value-Added
Effects	b	a	C	b	a	aC	a
	bC	aC		C	b	b	b
				bC	bC		C
				a			
Mean Squares	14.42	22.23	10.91	7802.49	32.64	57.29	65.59
	3.78	4.75		7560.42	52.08	76.86	163.52
				1343.36	23.29		87.37
				1078.84			
Degrees of Freedom	1,3074	3,3074	3,3074	1,3074	3,3074	9,3074	3,3074
	3,3074	9,3074		3,3074	1,3074	1,3074	1,3074
				3,3074	3,3074		3,3074
				3,3074			
F-Ratio	14.42	15.38	2.26	29.28	5.17	4.36	3.98
	3.78	3.28		28.37	8.24	5.85	9.93
				5.04	3.69		5.30
				4.05			
Significance	<.0002	<.0001	<.0788	<.0001	<.0015	<.0001	<.0076
	<.0101	<.0006		<.0001	<.0041	<.0156	<.0017
				<.0018	<.0114		<.0012
Least Square Estimates	-.75	+.15	-.34	-4.73	-.68	+.28	N.A.
	+.87	+.07	-.18	+6.86	-.02	+.90	
	+1.09	+.06	-.16	+8.54	-.04	+.48	
	+.82	-.41		+6.04	-.51	+2.82	
		-.23		-8.98	+1.10	+.92	
		.16		-7.98	+.64	-.02	
		1.12		-7.08	+.60	+2.10	
		.16		+5.50		+1.09	
		.28		-.54		+.34	
		.81		+1.48		-.52	
		-.00					
		-.01					

1950 and 1950-1960

Manufacturing Change Variable	All Establishments	Large Establishments	Medium Size Establishments	Employment Place of Residence	Employment Place of Work	Payroll	Value-Added
					Effects		
	BbC	Bb	Bb	Bb	Bb	C	Aa
		C		Aa	Aa	Aa	C
				C		AaC	Bb
				Mean Squares			
	19.68	63.45	27.05	7769.23	45.85	97.66	56.06
		12.42		3294.61	14.40	28.30	52.73
				1356.55		23.26	60.22
				Degrees of Freedom			
	6,2959	2,2959	2,2957	3,2959	2,2959	3,2959	6_1
		3,2959		2,2959	6,2959	6,2959	3_1
				6,2959		18,2959	2_1
				F-Ratio			
	2.04	44.04	5.45	29.91	7.30	7.59	3.41
		8.62		12.68	2.29	2.20	3.21
				5.22		1.81	3.66
				Significance			
	<.0576	<.0001	<.0044	<.0001	<.0007	<.0001	<.0024
		<.0001		<.0001	<.0327	<.0403	<.0221
				<.0001		<.0195	<.0258
				Least Square Estimates			
	+1.14	-.94	-.68	+10.32	-.77	+.88	-.44
	1.54	-.66	-.43	+4.97	-.48	-.97	+.44
	2.07	-.03		+2.91	-.49	-.64	-1.03
	.91	-.37		+2.51	+.28	-1.29	+1.55
	1.34	-.30		+5.76	+.14	+1.00	+.17
	.90			-9.94	+.44	-1.93	+.01
				-.23	+.11	+1.15	-.37
				+8.43	+.29	+.10	-.61
				+6.74		-.52	-.36
				+2.55		-4.61	-1.13
						+3.18	+.35
						-2.39	
						+1.25	
						-.97	
						-1.60	
						-4.23	
						+1.39	
						+1.28	
						-2.94	
						-1.56	
						-3.69	
						-.36	
						-.23	
						-.51	
						-.20	
						+1.68	
						+1.82	

of the C and BC effects on large plants can be interpreted as that of the increasing importance of locational differences within the largest market areas. Most important, however, is the fact that initial accessibility is the prime factor, thus lending support to the hypothesis of initial advantage and cumulative causation as well as to Harris's work. This fact does not preclude hierarchical diffusion; dispersion and decentralization were possible because of the shift of the entire system to higher levels of relative location.

4. The three basic components of relative location exerted independent influences on the rate of change employment by place of residence and most interactions between pairs of relative location components were not statistically significant. Apparently, initial relative accessibility is the most powerful effect. When B is held constant, A is second, when B and A are held constant, C is still important. The fourth interaction effect (AB) is a weak one. It is also important to note that not only did the proportion of counties growing increase as a consequence of diffusion, dispersion and decentralization, but at the extremes of relative location the rates are greater than the national average in the periphery and less (or even decline) in the core areas.

5. The relative insignificance of relative location to the change in employment in manufacturing by place of work is reflected in the analysis of variance (see the F-ratios). The fact that the BC and AC interactions were the only significant ones, and the way in which they interact, indicates that having high relative accessibility in 1950 and being close to an FEA city was not a particular "asset" in terms of growth of manufacturing employment. Further, it even indicates a relative decline in the most accessible counties in 1950. Dispersion and decentralization affected the spread of change in manufacturing employment more than they did the change in the number of manufacturing establishments.

6. The change in payroll in manufacturing and the change in value added provided different and somewhat contrasting results. In the case of the former, all the main effects and the two-way interaction effects were statistically significant. The single factor that most affected the degree of change in manufacturing payroll was relative location within urban fields. On the other hand, all main and interaction effects were weak with respect to change in manufacturing value added. Contrary to expectations, dispersion is not very important. Hierarchical diffusion is the most important process, and only secondly comes the relative location in urban fields. It seems that high value added industry decentralized to smaller urban centers and to counties in proximity to large metropolitan areas, and to areas that had

achieved thresholds for the support of manufacturing plants. Yet the dominance of the core manufacturing areas in terms of change in value added also remained unquestioned.

7. The results of the analysis of variance which are based on relative location in 1960 are not significantly different from those based on the relative location in 1950. That is, in spite of the important changes in relative location during the 1950's, the same conclusions continue to hold. This is an indication that the spatial changes unfolding were primarily evolutionary in nature.

B. For Change in Relative Location 1950-1960 and Rate of Change Differentials

1. The primary effect was the change in relative accessibility of counties within the nation, and the interaction of that change with location within urban fields. For example, the greater the change in population potential, the higher the rate of change in the number of all manufacturing establishments. But in contrast, the factor that affected the change in the total number of large manufacturing establishments most was the dynamics of population change in the central city of the FEA. It is interesting to note that rapid population change is negatively associated with the change in the number of large plants.

2. Change in relative location was strongly associated with the change in manufacturing employment by place of residence, with dispersion and decentralization the main factors. The A_i subeffects indicate a familiar pattern however: where the FEA city population grew most rapidly during the 1950's, the increase in manufacturing employment by place of residence tended to be relatively less than counties with FEA cities that grew "normally" or did not change. Growing cities suburbanized most rapidly, to put the matter most simply.

3. The association of change in relative location with the change in payroll and value-added in manufacturing provides an indication of the association between high wages, salaries, and value-added and rapid urbanization and industrialization.

C. For Initial Advantage and Change in Relative Location and Rate of Change Differentials

1. The most important conclusion is that for all variables except change in manufacturing payroll and in manufacturing value added, the most important factors associated with industrial change were initial relative accessibility and the change in such accessibility. That is, the basic factors contributing to an understanding of manufacturing change during the 1950's are initial

relative accessibility and its change. Dispersion was the most effective process influencing industrial change during the 1950's. Decentralization also was significant in a secondary sense for five of the seven industrial change variables, being most strongly associated with the change in the number of large establishments, in employment by place of residence and in payroll and value-added by manufacturing. Hierarchical diffusion played a similar secondary role.

REFERENCES

ALONSO, W. (1971), "The Economics of Urban Size," PAPERS OF THE REGIONAL SCIENCE ASSOCIATION, 26: 67-83.

ANDERSON, L. (1964), "Trickling Down: The Relationship Between Economic Growth and the Extent of Poverty among American Families," QUARTERLY JOURNAL OF ECONOMICS, 78: 511-523.

ARMSTRONG, R.B. (1972), THE OFFICE INDUSTRY. New York: Regional Plan Association.

BANFIELD, E.C. (1968), THE UNHEAVENLY CITY. Boston: Little, Brown.

BERGSMAN, J., P. GREENSTON, and R. HEALY. (1972), RESEARCH ON URBAN ECONOMIC DEVELOPMENT AND GROWTH POLICY. URBAN INSTITUTE PAPER 200-5. Washington, D.C: The Urban Institute.

———— (1972), "Determinants of Metropolitan Growth: 1965-1970," Paper presented at the Regional Science Association, Philadelphia, November, 1972.

BERRY, B.J.L. (1970), "The Geography of the United States in the Year 2000," TRANSACTIONS OF THE INSTITUTE OF BRITISH GEOGRAPHERS 51.

BORTS, G.H. and J.L. STEIN (1964), ECONOMIC GROWTH IN A FREE MARKET. New York: Columbia University Press.

BRECKENFELD, G. (1972), "Downtown has Fled to the Suburbs," FORTUNE, October.

BURROWS, J.C., C.E.METCALF, and J.B.KALER. (1971), INDUSTRIAL LOCATION IN THE UNITED STATES. Lexington: Heath Lexington Books.

CASSIDY, R. (1972), "Moving to the Suburbs," NEW REPUBLIC, January 22.

COHEN, Y.S. (1972), "Diffusion of an Innovation in an Urban System," UNIVERSITY OF CHICAGO DEPARTMENT OF GEOGRAPHY RESEARCH PAPER 140.

COMMISSION ON POPULATION GROWTH AND THE AMERICAN FUTURE (1972) POPULATION AND THE AMERICAN FUTURE. Washington, U.S.Government Printing Office.

CREAMER, D. (1943), "Shifts of Manufacturing Industries," in National Resources Planning Board, INDUSTRIAL LOCATION AND NATIONAL RESOURCES. Washington, D.C: U.S.Government Printing Office.

———— (1969), MANUFACTURING EMPLOYMENT BY TYPE OF LOCATION. Washington, D.C:The Conference Board.

DUE, J.F. (1961), "Studies of State-Local Tax Influences on Location of Industry," NATIONAL TAX JOURNAL, 14:167.

EBERHARD, J. (1966), "Technology for the City," INTERNATIONAL SCIENCE AND TECHNOLOGY, 57: 18-31.

FLOYD, J.S. (1952), EFFECTS OF TAXES ON INDUSTRIAL LOCATION. Chapel Hill: University of North Carolina Press.

FUCHS, V.R. (1962), CHANGES IN THE LOCATION OF MANUFACTURING IN THE UNITED STATES SINCE 1929. New Haven: Yale University Press.

GOODING, J. (1972), "Roadblocks Ahead for the Great Corporate Move-Out," FORTUNE, June.

HANDLIN, O. and J. BURCHARD(eds.) (1963) THE HISTORIAN AND THE CITY. Cambridge, Mass: MIT Press.

HARRIS, C.C., and F.E.HOPKINS (1972), LOCATIONAL ANALYSIS: AN INTERREGIONAL ECONOMETRIC MODEL OF AGRICULTURE, MINING, MANUFACTURING, AND SERVICES. Lexington: Heath Lexington Books.

HARRIS, C.D. (1954), "The Market as a Factor in the Localization of Industry in the United States," ANNALS OF THE ASSOCIATION OF AMERICAN GEOGRAPHERS, 64:315-348.

HARTNETT, H.D. (1971), " A Locational Analysis of Those Manufacturing Firms That Have Located and Relocated within the City of Chicago, 1955-1968,"

HERR, P.B. (1960), "The Regional Impact of Highways," URBAN LAND, 19: 3-8.

JAMES, F.J,Jr., and J.W.HUGHES. (1973), "The Process of Employment Location Change: An Empirical Analysis," LAND ECONOMICS, November.

KNEELAND, D.E. (1972), "Quiet Decay Erodes Downtown Areas of Small Cities, " NEW YORK TIMES, February 8.

LEONE, R.A. (1972), "The Role of Data Availability in Intrametropolitan Workplace Location Studies," ANNALS OF ECONOMIC AND SOCIAL MEASUREMENT 1:171-182.

LICHTENBERG, R.M. (1960), ONE TENTH OF A NATION. Cambridge, Mass.: Harvard University Press.

MADDEN, J.L. (1970), METROPOLITAN GROWTH CENTERS IN HISTORICAL PERSPECTIVE. Discussion Paper 35, Growth Center Research Project, University of Kentucky, Lexington.

MELTZER, J. (1972), "Loop Betrays Deeper Ills of City," CHICAGO SUN-TIMES VIEWPOINT, October 22.

MILLS, E.S. (1972), URBAN ECONOMICS. Chicago: Scott, Foresman.

MUTH, R.F. (1968), "Differential Growth Among Large U.S. Cities," In PAPERS IN QUANTITATIVE ECONOMICS, J.P. Quirk and A.M.Zarley, eds. Lawrence: University of Kansas Press, 315-355.

NEWMAN, D.K. (1967), "Decentralization of Jobs," MONTHLY LABOR REVIEW 90: 7-13.

OLSEN, R.J. and G.W.WESTLEY (1973), "Regional Differences in the Growth of Market Potentials, 1950-1970," Paper presented at the Mid-Continent Regional Science Association Meetin Stillwater, Oklahoma, April 1973.

______ and ______ (1973) ,"Regional Differences in the Growth of Overnight Truck Transport Market Areas, 1950-1970," Paper presented at the Southern Regional Science Associations Meetings, New Orleans, La., April 1973.

______, ______ and L.G.BRAY (1974), "The Location of Manufacturing Employment in BEA Economic Areas: A Regression Analysis Using 1950, 1960, and 1970 Data," Paper presented at the Southern Regional Science Association Meetings, Rosslyn, Virginia, April 1974.

REDER, M.W. (1955), "The Theory of Occupational Wage Differentials," AMERICAN ECONOMIC REVIEW 45:832-852.

ROTERUS, V. (1970), "Do Small Towns Have a Future?" URBAN LAND, March: 12-15.

SIEBERT, H. (1965), REGIONAL ECONOMIC GROWTH: THEORY AND POLICY. Scranton, Pa.: International Textbook Company.

SMOLENSKY, E. (1965) , "Poverty Policy and Pennsylvania," Paper presented at the meeting of the Pennsylvania Economic Association.

SPIEGELMAN, R.G. (1968), A STUDY OF INDUSTRY LOCATION USING MULTIPLE REGRESSION TECHNIQUES. AGRICULTURAL ECONOMICS REPORT 140, Economic Research Service, U.S.Department of Agriculture, 1968.

STANBACK, T.M. and R.V.KNIGHT (1970), THE METROPOLITAN ECONOMY. New York: Columbia University Press.

STANFORD RESEARCH INSTITUTE (1968), COSTS OF URBAN INFRASTRUCTURE AS RELATED TO CITY SIZE IN DEVELOPING COUNTRIES. Palo Alto, California: The Institute.

STEVENS, B.H., R.C.DOUGLAS, and C.B.NEIGHBOR (1969),Trends in Industrial Location and their impact on Regional Economic Development," Philadelphia: Regional Science Research Institute Discussion Paper No. 31.

THOMPSON, W. (1968), "Internal and External Facots in the Development of Urban Economics, " In ISSUES IN URBAN ECONOMICS, H.S.Perloff and L. Wingo, eds. Baltimore: The Johns Hopkins University Press.

TIDEMAN, T.N. (1968), "The Theoretical Efficiency of 'Potential' and Transport Models of Location," Paper prepared at the Center for Urban Studies University of Chicago.

TILL, T. (1973), "The Extent of Industrialization in Southern Nonmetro Labor Markets in the 1960's," JOURNAL OF REGIONAL SCIENCE 13:453-461.

U.S. DEPARTMENT OF COMMERCE (1959), METROPOLITAN AREA AND CITY SIZE PATTERNS OF MANUFACTURING INDUSTRIES, 1954. Washington: U.S.Government Printing Office.

WEISS, L.W. (1969), "The Geographical Size of Markets in Manufacturing," Paper issued in the Social Systems Research Institute Workshop Series, University of Wisconsin.

WONNACOTT, R.J. (1963), MANUFACTURING COSTS AND THE COMPARATIVE ADVANTAGE OF UNITE STATES REGIONS. Minneapolis: Upper Midwest Research and Development Conncil and the University of Minnesota.

YOUNG, D. (1972), "Industry Flees Decaying City," CHICAGO TRIBUNE, May 7.

THE UNIVERSITY OF CHICAGO
DEPARTMENT OF GEOGRAPHY
RESEARCH PAPERS (Lithographed, 6×9 Inches)

(Available from Department of Geography, The University of Chicago, 5828 S. University Ave., Chicago, Illinois 60637. Price: $6.00 each; by series subscription, $5.00 each.)

106. SAARINEN, THOMAS F. *Perception of the Drought Hazard on the Great Plains* 1966. 183 pp.
107. SOLZMAN, DAVID M. *Waterway Industrial Sites: A Chicago Case Study* 1967. 138 pp.
108. KASPERSON, ROGER E. *The Dodecanese: Diversity and Unity in Island Politics* 1967. 184 pp.
109. LOWENTHAL, DAVID, et al. *Environmental Perception and Behavior.* 1967. 88 pp.
110. REED, WALLACE E. *Areal Interaction in India: Commodity Flows of the Bengal-Bihar Industrial Area* 1967. 210 pp.
112. BOURNE, LARRY S. *Private Redevelopment of the Central City: Spatial Processes of Structural Change in the City of Toronto* 1967. 199 pp.
113. BRUSH, JOHN E., and GAUTHIER, HOWARD L., JR. *Service Centers and Consumer Trips: Studies on the Philadelphia Metropolitan Fringe* 1968. 182 pp.
114. CLARKSON, JAMES D. *The Cultural Ecology of a Chinese Village: Cameron Highlands, Malaysia* 1968. 174 pp.
115. BURTON, IAN; KATES, ROBERT W.; and SNEAD, RODMAN E. *The Human Ecology of Coastal Flood Hazard in Megalopolis* 1968. 196 pp.
117. WONG, SHUE TUCK. *Perception of Choice and Factors Affecting Industrial Water Supply Decisions in Northeastern Illinois* 1968. 96 pp.
118. JOHNSON, DOUGLAS L. *The Nature of Nomadism* 1969. 200 pp.
119. DIENES, LESLIE. *Locational Factors and Locational Developments in the Soviet Chemical Industry* 1969. 285 pp.
120. MIHELIC, DUSAN. *The Political Element in the Port Geography of Trieste* 1969. 104 pp.
121. BAUMANN, DUANE. *The Recreational Use of Domestic Water Supply Reservoirs: Perception and Choice* 1969. 125 pp.
122. LIND, AULIS O. *Coastal Landforms of Cat Island, Bahamas: A Study of Holocene Accretionary Topography and Sea-Level Change* 1969. 156 pp.
123. WHITNEY, JOSEPH. *China: Area, Administration and Nation Building* 1970. 198 pp.
124. EARICKSON, ROBERT. *The Spatial Behavior of Hospital Patients: A Behavioral Approach to Spatial Interaction in Metropolitan Chicago* 1970. 198 pp.
125. DAY, JOHN C. *Managing the Lower Rio Grande: An Experience in International River Development* 1970. 277 pp.
126. MAC IVER, IAN. *Urban Water Supply Alternatives: Perception and Choice in the Grand Basin, Ontario* 1970. 178 pp.
127. GOHEEN, PETER G. *Victorian Toronto, 1850 to 1900: Pattern and Process of Growth* 1970. 278 pp.
128. GOOD, CHARLES M. *Rural Markets and Trade in East Africa* 1970. 252 pp.
129. MEYER, DAVID R. *Spatial Variation of Black Urban Households* 1970. 127 pp.
130. GLADFELTER, BRUCE. *Meseta and Campiña Landforms in Central Spain: A Geomorphology of the Alto Henares Basin* 1971. 204 pp.
131. NEILS, ELAINE M. *Reservation to City: Indian Urbanization and Federal Relocation* 1971. 200 pp.
132. MOLINE, NORMAN T. *Mobility and the Small Town, 1900–1930* 1971. 169 pp.
133. SCHWIND, PAUL J. *Migration and Regional Development in the United States, 1950–1960* 1971. 170 pp.
134. PYLE, GERALD F. *Heart Disease, Cancer and Stroke in Chicago: A Geographical Analysis with Facilities Plans for 1980* 1971. 292 pp.
135. JOHNSON, JAMES F. *Renovated Waste Water: An Alternative Source of Municipal Water Supply in the U.S.* 1971. 155 pp.
136. BUTZER, KARL W. *Recent History of an Ethiopian Delta: The Omo River and the Level of Lake Rudolf* 1971. 184 pp.
137. HARRIS, CHAUNCY D. *Annotated World List of Selected Current Geographical Serials in English, French, and German* 3rd edition 1971. 77 pp.
138. HARRIS, CHAUNCY D., and FELLMANN, JEROME D. *International List of Geographical Serials* 2nd edition 1971. 267 pp.
139. MC MANIS, DOUGLAS R. *European Impressions of the New England Coast, 1497–1620* 1972. 147 pp.
140. COHEN, YEHOSHUA S. *Diffusion of an Innovation in an Urban System: The Spread of Planned Regional Shopping Centers in the United States, 1949–1968* 1972. 136 pp.

141. MITCHELL, NORA. *The Indian Hill-Station: Kodaikanal* 1972. 199 pp.

142. PLATT, RUTHERFORD H. *The Open Space Decision Process: Spatial Allocation of Costs and Benefits* 1972. 189 pp.

143. GOLANT, STEPHEN M. *The Residential Location and Spatial Behavior of the Elderly: A Canadian Example* 1972. 226 pp.

144. PANNELL, CLIFTON W. *T'ai-chung, T'ai-wan: Structure and Function* 1973. 200 pp.

145. LANKFORD, PHILIP M. *Regional Incomes in the United States, 1929–1967: Level, Distribution, Stability, and Growth* 1972. 137 pp.

146. FREEMAN, DONALD B. *International Trade, Migration, and Capital Flows: A Quantitative Analysis of Spatial Economic Interaction* 1973. 202 pp.

147. MYERS, SARAH K. *Language Shift Among Migrants to Lima, Peru* 1973. 204 pp.

148. JOHNSON, DOUGLAS L. *Jabal al-Akhdar, Cyrenaica: An Historical Geography of Settlement and Livelihood* 1973. 240 pp.

149. YEUNG, YUE-MAN. *National Development Policy and Urban Transformation in Singapore: A Study of Public Housing and the Marketing System* 1973. 204 pp.

150. HALL, FRED L. *Location Criteria for High Schools: Student Transportation and Racial Integration* 1973. 156 pp.

151. ROSENBERG, TERRY J. *Residence, Employment, and Mobility of Puerto Ricans in New York City* 1974. 230 pp.

152. MIKESELL, MARVIN W., editor. *Geographers Abroad: Essays on the Problems and Prospects of Research in Foreign Areas* 1973. 296 pp.

153. OSBORN, JAMES. *Area, Development Policy, and the Middle City in Malaysia* 1974. 273 pp.

154. WACHT, WALTER F. *The Domestic Air Transportation Network of the United States* 1974. 98 pp.

155. BERRY, BRIAN J. L., et al. *Land Use, Urban Form and Environmental Quality* 1974. 464 pp.

156. MITCHELL, JAMES K. *Community Response to Coastal Erosion: Individual and Collective Adjustments to Hazard on the Atlantic Shore* 1974. 209 pp.

157. COOK, GILLIAN P. *Spatial Dynamics of Business Growth in the Witwatersrand* 1975. 143 pp.

158. STARR, JOHN T., JR. *The Evolution of Unit Train Operations in the United States: 1960–1969—A Decade of Experience* 1975.

159. PYLE, GERALD F. *The Spatial Dynamics of Crime* 1974. 220 pp.

160. MEYER, JUDITH W. *Diffusion of an American Montessori Education* 1975. 109 pp.

161. SCHMID, JAMES A. *Urban Vegetation: A Review and Chicago Case Study* 1975. 280 pp.

162. LAMB, RICHARD. *Metropolitan Impacts on Rural America* 1975. 210 pp.

163. FEDOR, THOMAS. *Patterns of Urban Growth in the Russian Empire during the Nineteenth Century* 1975. 275 pp.

164. HARRIS, CHAUNCY D. *Guide to Geographical Bibliographies and Reference Works in Russian or on the Soviet Union* 1975. 496 pp.

165. JONES, DONALD W. *Migration and Urban Unemployment in Dualistic Economic Development* 1975. 186 pp.

166. BEDNARZ, ROBERT S. *The Effect of Air Pollution on Property Value* 1975. 118 pp.

167. HANNEMANN, MANFRED. *The Diffusion of the Reformation in Southwestern Germany, 1518-1534* 1975. 248 pp.

168. SUBLETT, MICHAEL D. *Farmers on the Road. Interfarm Migration and the Farming of Noncontiguous Lands in Three Midwestern Townships, 1939-1969* 1975. 228 pp.

169. STETZER, DONALD FOSTER. *Special Districts in Cook County: Toward a Geography of Local Government* 1975. 189 pp.

170. EARLE, CARVILLE V. *The Evolution of a Tidewater Settlement System: All Hallow's Parish, Maryland, 1650–1783* 1975. 249 pp.

171. SPODEK, HOWARD. *Urban-Rural Integration in Regional Development: A Case Study of Saurashtra, India—1800–1960* 1975.

172. COHEN, YEHOSHUA S. and BERRY, BRIAN J. L. *Spatial Components of Manufacturing Change* 1975. 272 pp.

173. HAYES, CHARLES R. *The Dispersed City: The Case of Piedmont, North Carolina* 1975.

174. CARGO, DOUGLAS B. *Solid Wastes: Factors Influencing Generation Rates* 1975.

175. GILLARD, QUENTIN. *Incomes and Accessibility. Metropolitan Labor Force Participation, Commuting, and Income Differentials in the United States, 1960–1970* 1975.

176. MORGAN, DAVID J. *Patterns of Population Distribution: A Residential Preference Model and Its Dynamic* 1975.

177. STOKES, HOUSTON H.; JONES, DONALD W. and NEUBURGER, HUGH M. *Unemployment and Adjustment in the Labor Market: A Comparison between the Regional and National Responses* 1975. 135 pp.